A HISTORY OF π (PI)

A HISTORY OF π (PI)

Petr Beckmann

Electrical Engineering Department
University of Colorado

BARNES
& NOBLE
BOOKS
NEW YORK

TO ŠMUDLA,
who never doubted the success of this book

This edition published by Barnes & Noble, Inc.,
by arrangement with St. Martin's Press, Inc.

1993 Barnes & Noble Books

ISBN 0-88029-418-3

Printed and bound in the United States of America

M 20 19 18 17

Preface

The history of π is a quaint little mirror of the history of man. It is the story of men like Archimedes of Syracuse, whose method of calculating π defied substantial improvement for some 1900 years; and it is also the story of a Cleveland businessman, who published a book in 1931 announcing the grand discovery that π was exactly equal to 256/81, a value that the Egyptians had used some 4,000 years ago. It is the story of human achievement at the University of Alexandria in the 3rd century B.C.; and it is also the story of human folly which made mediaeval bishops and crusaders set the torch to scientific libraries because they condemned their contents as works of the devil.

Being neither an historian nor a mathematician, I felt eminently qualified to write that story.

That remark is meant to be sarcastic, but there is a kernel of truth in it. Not being an historian, I am not obliged to wear the mask of dispassionate aloofness. History relates of certain men and institutions that I admire, and others that I detest; and in neither case have I hesitated to give vent to my opinions. However, I believe that facts and opinion are clearly separated in the following, so that the reader should run no risk of being overly influenced by my tastes and prejudices.

Not being a mathematician, I am not obliged to complicate my explanations by excessive mathematical rigor. It is my hope that this little book might stimulate non-mathematical readers to become interested in mathematics, just as it is my hope that students of physics and engineering might become interested in the history of the tools they are using in their work. There are, however, two sure and all too well tried methods of how to make mathematics repugnant: One is to brutalize the reader by assertions without proof; the other is to hit him over the head with epsilonics and proofs of existence and unicity. I have tried to steer a middle course between the two.

A history of π containing only the bare facts and dates when who did what to π tends to be rather dull, and I thought it more interesting to mix in some of the background of the times in which π made progress. Sometimes I have strayed rather far afield, as in the case of the Roman Empire and the Middle Ages; but I thought it just as important to explore the times when π did *not* make any progress, and why it did not make any.

The mathematical level of the book is flexible. The reader who finds the mathematics too difficult in some places is urged to do what the mathematician will do when he finds it too trivial: Skip it.

This book, small as it is, would not have been possible without the wholehearted cooperation of the staff of Golem Press, and I take this opportunity to express my gratitude to every one of them. I am also indebted to the Archives Division of the Indiana State Library for making available photostats of Bill 246, Indiana House of Representatives, 1897, and to the Cambridge University Press, Dover Publications and Litton Industries for granting permission to reproduce copyrighted materials without charge. Their courtesy is acknowledged in the notes accompanying the individual figures.

I much enjoyed writing this book, and it is my sincere hope that the reader will enjoy reading it, too.

Boulder, Colorado
August 1970

Petr Beckmann

PREFACE TO THE SECOND EDITION

After all but calling Aristotle a dunce, spitting on the Roman Empire, and flipping my nose at some other highly esteemed institutions, I had braced myself for the reviews that would call this book the sick product of an insolent ignoramus. My surprise was therefore all the more pleasant when the reviews were very favorable, and the first edition went out of print in less than a year.

I am most grateful to the many readers who have written in to point out misprints and errors, particularly to those who took me to task (quite rightly) for ignoring the recent history of evaluating π by digital computers. I have attempted to remedy this shortcoming by adding a chapter on π in the computer age.

Mr. D.S. Candelaber of Golem Press had the bright idea of imprinting the end sheets of the book with the first 10,000 decimal places of π, and the American Mathematical Society kindly gave permission to reproduce the first two pages of the computer print-out as published by Shanks and Wrench in 1962. A reprint of this work was very kindly made available by one of the authors, Dr. John W. Wrench, Jr. To all of these, I would like to express my sincere thanks. I am also most grateful to all readers who have given me the benefit of their comments. I am particularly indebted to Mr. Craige Schensted of Ann Arbor, Michigan, and M. Jean Meeus of Erps-Kwerps, Belgium, for their detailed lists of misprints and errors in the first edition.

Boulder, Colorado *P.B.*
May 1971

PREFACE TO THE THIRD EDITION

Some more errors have been corrected and the type has been re-set for the third edition.

A Japanese translation of this book was published in 1973.

Meanwhile, a disturbing trend away from science and toward the irrational has set in. The aerospace industry has been all but dismantled. College enrollment in the hard sciences and engineering has significantly dropped. The disoriented and the gullible flock in droves to the various Maharajas of Mumbo Jumbo. Ecology, once a respected scientific discipline, has become the buzzword of frustrated housewives on messianic ego-trips. Technology has wounded affluent intellectuals with the ultimate insult: They cannot understand it any more.

Ignorance, anti-scientific and anti-technology sentiment have always provided the breeding ground for tyrannies in the past. The power of the ancient emperors, the mediaeval Church, the Sun Kings, the State with a capital S, was always rooted in the ignorance of the oppressed. Anti-scientific and anti-technology sentiment is providing a breeding ground for encroaching on the individual's freedoms now. A new tyranny is on the horizon. It masquerades under the meaningless name of "Society."

Those who have not learned the lessons of history are destined to relive it.

Must the rest of us relive it, too?

Boulder, Colorado *P.B.*
Christmas 1974

Contents

1

DAWN

History records the names of royal bastards, but cannot tell us the origin of wheat.

JEAN HENRI FABRE
(1823-1915)

A million years or so have passed since the tool-wielding animal called man made its appearance on this planet. During this time it learned to recognize shapes and directions; to grasp the concepts of magnitude and number; to measure; and to realize that there exist relationships between certain magnitudes.

The details of this process are unknown. The first dim flash in the darkness goes back to the stone age — the bone of a wolf with incisions to form a tally stick (see figure on next page). The flashes become brighter and more numerous as time goes on, but not until about 2,000 B.C. do the hard facts start to emerge by direct documentation rather than by circumstantial evidence. And one of these facts is this: By 2,000 B.C., men had grasped the significance of the constant that is today denoted by π, and that they had found a rough approximation of its value.

How had they arrived at this point? To answer this question, we must return into the stone age and beyond, and into the realm of speculation.

Long before the invention of the wheel, man must have learned to identify the peculiarly regular shape of the circle. He saw it in the pupils of his fellow men and fellow animals; he saw it bounding the disks of the Moon and Sun; he saw it, or something near it, in some flowers; and perhaps he was pleased by its infinite symmetry as he drew its shape in the sand with a stick.

Then, one might speculate, men began to grasp the concept of magnitude — there were large circles and small circles, tall trees and small trees, heavy stones, heavier stones, very heavy stones. The transition from these qualitative statements to quantitative measurement was the dawn of mathematics. It must have been a long and arduous road, but it is a safe guess that it was first taken for quantities that assume only integral values — people, animals, trees, stones, sticks. For counting is a quantitative measurement: The measurement of the amount of a multitude of items.

Man first learned to count to two, and a long time elapsed before he learned to count to higher numbers. There is a fair amount of evidence for this,[1] perhaps none of it more fascinating than that preserved in man's languages: In Czech, until the Middle Ages, there used to be two kinds of plural — one for two items, another for many (more than two) items, and apparently in Finnish this is so to this day. There is evidently no connection between the (Germanic) words *two* and *half*; there is none in the Romance languages (French: *deux* and *moitié*) nor in the Slavic languages (Russian: *dva* and *pol*), and in Hungarian, which is not an Indo-European language, the words are *kettő* and *fél*. Yet in all European languages, the words for 3 and 1/3, 4 and 1/4, etc., are related. This suggests that men grasped the concept of a ratio, and the idea of a relation between a number and its reciprocal, only after they had learned to count beyond two.

A stone age tally stick. The tibia (shin) of a wolf with two long incisions in the center, and two series of 25 and 30 marks. Found in Věstonice, Moravia (Czechoslovakia) in 1937.[2]

The next step was to discover relations between various magnitudes. Again, it seems certain that such relations were first expressed qualitatively. It must have been noticed that bigger stones are heavier, or to put it into more complicated words, that there is a relation between the volume and the weight of a stone. It must have been observed that an older tree is taller, that a faster runner covers a longer distance, that more prey gives more food, that larger fields yield bigger crops. Among all these kinds of relationships, there was one which could hardly have escaped notice, and which, moreover, had no exceptions:

The wider a circle is "across," the longer it is "around."

And again, this line of qualitative reasoning must have been followed by quantitative considerations. If the volume of a stone is doubled, the weight is doubled; if you run twice as fast, you cover double the distance; if you treble the fields, you treble the crop; if you double the diameter of a circle, you double its circumference. Of course, the rule does not always work: A tree twice as old is not twice as tall. The reason is that "the more . . . the more" does not always imply proportionality; or in more snobbish words, not every monotonic function is linear.

Neolithic man was hardly concerned with monotonic functions; but it is certain that men learned to recognize, consciously or unconsciously, by experience, instinct, reasoning, or all of these, the concept of proportionality; that is, they learned to recognize pairs of magnitude such that if the one was doubled, trebled, quadrupled, halved or left alone, then the other would also double, treble, quadruple, halve or show no change.

And then came the great discovery. By recognizing certain specific properties, and by defining them, little is accomplished. (That is why the old type of descriptive biology was so barren.) But a great scientific discovery has been made when the observations are generalized in such a way that a generally valid rule can be stated. The greater its range of validity, the greater its significance. To say that one field will feed half the tribe, two fields will field the whole tribe, three fields will feed one and a half tribes, all this applies only to certain fields and tribes. To say that one bee has six legs, three bees have eighteen legs, etc., is a statement that applies, at best, to the class of insects.But somewhere along the line some inquisitive and smart individuals must have seen something in common in the behavior of the magnitudes in these and similar statements:

No matter how the two proportional quantities are varied, their ratio remains constant.

For the fields, this constant is $1 : \frac{1}{2} = 2 : 1 = 3 : 1\frac{1}{2} = 2$. For the bees, this constant is $1 : 6 = 3 : 18 = 1/6$. And thus, man had discovered a general, not a specific, truth.

This constant ratio was not obtained by numerical division (and certainly not by the use of Arabic numerals, as above); more likely, the ratio was expressed geometrically, for geometry was the first mathematical discipline to make substantial progress. But the actual technique of arriving at the constancy of the ratio of two proportional quantities makes little difference to the argument.

There were of course many intermediate steps, such as the discovery of sums, differences, products and ratios; and the step of abstraction, exemplified by the transition from the statement "two birds and two birds make four birds" to the statement "two and two is four." But the decisive and great step on the road to π was the discovery that proportional quantities have a constant ratio.

From here it was but a dwarf's step to the constant π: If the "around" (circumference) and the "across" (diameter) of a circle were recognized as proportional quantities, as they easily must have been, then it immediately follows that the ratio

circumference : diameter = constant for all circles.

This constant circle ratio was not denoted by the symbol π until the 18th century (A.D.), nor, for that matter, did the equal sign (=) come into general use before the 16th century A.D. (The twin lines as an equal sign were used by the English physician and mathematician Robert Recorde in 1557 with the charming explanation that "noe .2. thynges, can be moare equalle.") However, we shall use modern notation from the outset, so that the definition of the number π reads

$$\pi = \frac{C}{D} \tag{1}$$

where C is the circumference, and D the diameter of any circle.

And with this, our speculative road has reached, about 2,000 B.C., the dawn of the documented history of mathematics. From the documents of that time it is evident that by then the Babylonians and the Egyptians (at least) were aware of the existence and significance of the constant π as given by (1).

BUT the Babylonians and the Egyptians knew more about π than its mere existence. They had also found its approximate value. By about 2,000 B.C., the Babylonians had arrived at the value

$$\pi = 3\,{}^1\!/_8 \tag{2}$$

and the Egyptians at the value

$$\pi = 4(8/9)^2 \tag{3}$$

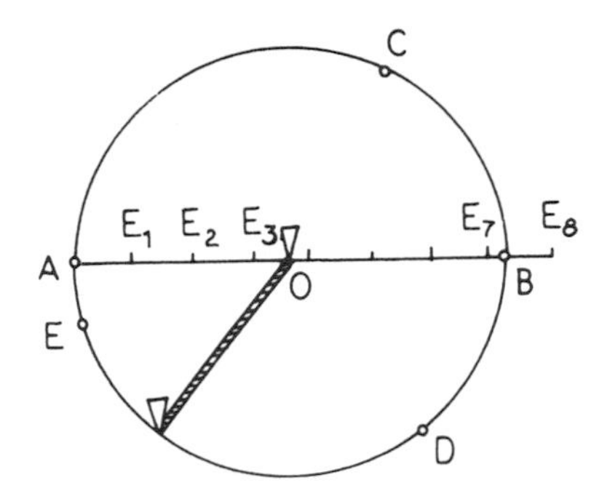

How to measure π in the sands of the Nile

How did these ancient people arrive at these values? Nobody knows for certain, but this time the guessing is fairly easy.

Obviously, the easiest way is to take a circle, to measure its circumference and diameter, and to find π as the ratio of the two. Let us try to do just that, imagining that we are in Egypt in 3,000 B.C. There is no National Bureau of Standards; no calibrated measuring tapes. We are not allowed to use the decimal system or numerical division of any kind. No compasses, no pencil, no paper; all we have is stakes, ropes and sand.

So we find a fairly flat patch of wet sand along the Nile, drive in a stake, attach a piece of rope to it by loop and knot, tie the other end to another stake with a sharp point, and keeping the rope taut, we draw a circle in the sand. We pull out the central stake, leaving a hole *O* (see figure above). Now we take a longer piece of rope, choose any point *A* on the circle and stretch the rope from *A* across the hole *O* until it intersects the circle at *B*. We mark the length *AB* on the rope (with charcoal); this is the diameter of the circle and our unit of length. Now we take the rope and lay it into the circular groove in the sand, starting at *A*. The charcoal mark is at *C*; we have laid off the diameter along the circumference once. Then we lay it off a second time from *C* to *D*, and a third time from *D* to *A*, so that the diameter goes into the circumference three (plus a little bit) times.

If, to start with, we neglect the little bit, we have, to the nearest integer,

$$\pi = 3 \tag{3}$$

To improve our approximation, we next measure the little left-over bit *EA* as a fraction of our unit distance *AB*. We measure the curved

23 וַיַּעַשׂ אֶת־הַיָּם מוּצָק עֶשֶׂר
בָּאַמָּה מִשְּׂפָתוֹ עַד־שְׂפָתוֹ עָגֹל ׀ סָבִיב וְחָמֵשׁ בָּאַמָּה
קוֹמָתוֹ וְקָוֹה שְׁלֹשִׁים בָּאַמָּה יָסֹב אֹתוֹ סָבִיב׃

23. και εποιησε την θαλασσαν δεκα εν πηχει απο του χειλους αυτης εως του ψειλους αυτης, στρογγυλον κυελω το αυτο. πεντε εν πηχει το υψος αυτης. και συνηγμενη τρεις και τριακοντα εν πιχει.

23 Hizo asimismo un mar de fundición, de diez codos del uno al otro lado, redondo, y de cinco codos de alto, y ceñíalo en derredor un cordón de treinta codos.

23. Il fit aussi une mer de fonte, de dix coudées d'un bord jusqu'à l'autre, qui était toute ronde: elle avait cinq coudées de haut, et elle était environnée tout à l'entour d'un cordon de trente coudées.

23. Udělal též moře slité, desíti loket od jednoho kraje k druhému, okrouhlé vůkol, a pět loket byla vysokost jeho, a okolek jeho třicet loket vůkol.

23. Und er machte ein Meer, gegossen, von einem Rand zum andern zehn Ellen weit, rundumher, und fünf Ellen hoch, und eine Schnur dreißig Ellen lang war das Maß ringsum.

23. And he made a molten sea, ten cubits from the one brim to the other; it was round all about, and his height was five cubits: and a line of thirty cubits did compass it round about.

length EA and mark it on a piece of rope. Then we straighten the rope and lay it off along AB as many times as it will go. It will go into our unit distance AB between 7 and 8 times. (Actually, if we swindle a little and check by 20th century arithmetic, we find that 7 is much nearer the right value than 8, i.e., that E_7 in the figure on p. 13 is nearer to B than E_8, for $1/7 = 0.142857...$, $1/8 = 0.125$, and the former value is nearer $\pi - 3 = 0.141592...$ However, that would be difficult to ascertain by our measurement using thick, elastic ropes with coarse charcoal marks for the roughly circular curve in the sand whose surface was judged "flat" by arbitrary opinion.)

We have thus measured the length of the arc EA to be between 1/7 and 1/8 of the unit distance AB; and our second approximation is therefore

$$3\tfrac{1}{8} < \pi < 3\tfrac{1}{7} \tag{4}$$

for this, to the nearest simple fractions, is how often the unit rope length AB goes into the circumference $ABCD$.

And indeed, the values

$$\pi = 3, \qquad \pi = 3\tfrac{1}{7}, \qquad \pi = 3\tfrac{1}{8}$$

are the values most often met in antiquity.

For example, in the Old Testament (I Kings vii.23, and 2 Chronicles iv.2), we find the following verse:

> *"Also, he made a molten sea of ten cubits from brim to brim, round in compass, and five cubits the height thereof; and a line of thirty cubits did compass it round about."*

The molten sea, we are told, is round; it measures 30 cubits round about (in circumference) and 10 cubits from brim to brim (in diameter); thus the biblical value of π is $30/10 = 3$.

The Book of Kings was edited by the ancient Jews as a religious work about 550 B.C., but its sources date back several centuries. At that time, π was already known to a considerably better accuracy, but evidently not to the editors of the Bible. The Jewish Talmud, which is essentially a commentary on the Old Testament, was published about 500 A.D. Even at this late date it also states "that which in circumference is three hands broad is one hand broad."

The molten sea as reconstructed by Gressman from the description in 2 Kings vii.[2]

In early antiquity, in Egypt and other places, the priests were often closely connected with mathematics (as custodians of the calendar, and for other reasons to be discussed later). But as the process of specialization in society continued, science and religion drifted apart. By the time the Old Testament was edited, the two were already separated. The inaccuracy of the biblical value of π is, of course, no more than an amusing curiosity. Nevertheless, with the hindsight of what happened afterwards, it is interesting to note this little pebble on the road to confrontation between science and religion, which on several occasions broke out into open conflict, and about which we shall have more to say later.

Returning to the determination of π by direct measurement using primitive equipment, it can probably safely be said that it led to values no better than (4).

From now on, man had to rely on his wits rather than on ropes and stakes in the sand. And it was by his wits, rather than by experimental measurement, that he found the circle's area.

THE ancient peoples had rules for calculating the area of a circle. Again, we do not know how they derived them (except for one method used in Egypt, to be described in the next chapter), and once

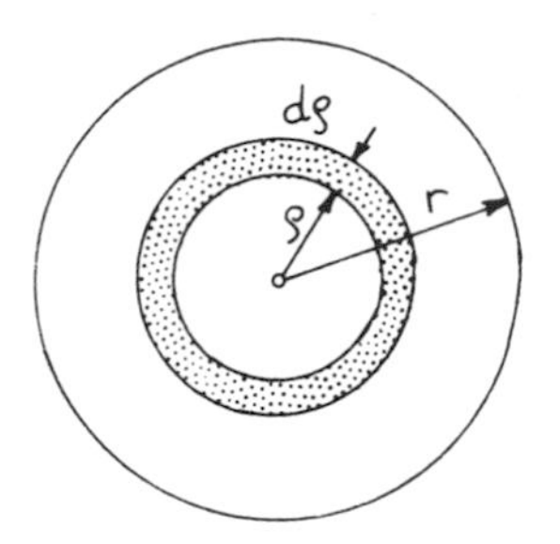

Calculation of the area of a circle by integral calculus. The area of an elementary ring is $dA = 2\pi\rho d\rho$; hence the area of the circle is

$$A = 2\pi \int_0^r \rho \, d\rho = \pi r^2.$$

more we have to play the game "How do you do it with their knowledge" to make a guess. The area of a circle, we know, is

$$A = \pi r^2 \tag{5}$$

where r is the radius of the circle. Most of us first learned this formula in school with the justification that teacher said so, take it or leave it, but you better take it and learn it by heart; the formula is, in fact, an example of the brutality with which mathematics is often taught to the innocent. Those who later take a course in the integral calculus learn that the derivation of (5) is quite easy (see figure above). But how did people calculate the area of a circle almost five millenia before the integral calculus was invented?

They probably did it by a method of rearrangement. They calculated the area of a rectangle as length times width. To calculate the area of a parallelogram, they could construct a rectangle of equal area by rearrangement as in the figure below, and thus they found that the area of a parallelogram is given by base times height. The age of rigor that came with the later Greeks was still far away; they

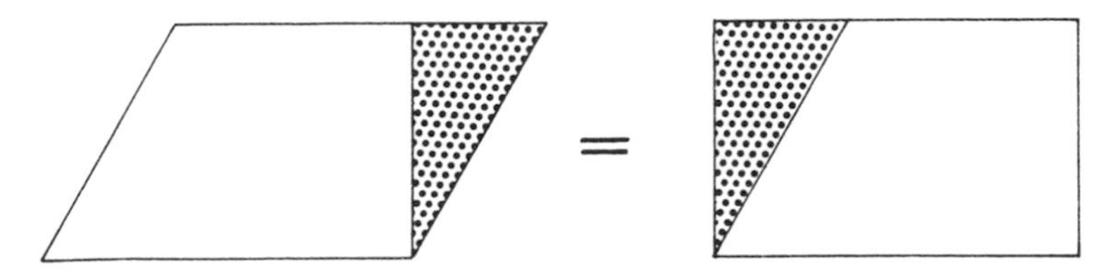

The parallelogram and the rectangle have equal areas, as seen by cutting off the shaded triangle and reinserting it as indicated.

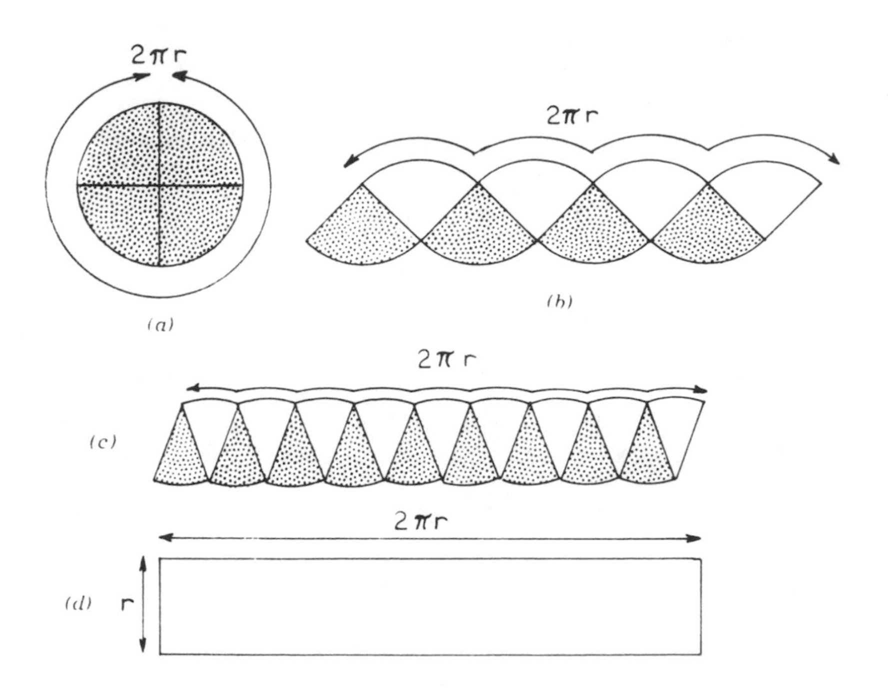

Determination of the area of a circle by rearrangement.
The areas of the figures (b), (c), (d) equal exactly double
the area of circle (a).

did not have to know about congruent triangles to be convinced by the "obvious" validity of the rearrangement.

So now let us try to use the general idea of rearrangement as in the figure above to convert a circle to a parallelogram of equal area. We are still using sticks to draw pictures in the sand, but this time we do this only to help our imagination, not to perform an actual measurement.

We first cut up a circle into four quadrants as in (a) above, and arrange them as shown in figure (b). Then we fill in the spaces between the segments by four equally large quadrants. The outline of the resulting weird figure is vaguely reminiscent of a parallelogram. The length of the figure, measured along the circular arcs, is equal to the circumference of the original circle, $2\pi r$. What we can say with certainty is that the area of this figure is exactly double the area of the original circle.

If we now divide the circle not into four, but into very many segments, our quasi-parallelogram (c) will resemble a parallelogram

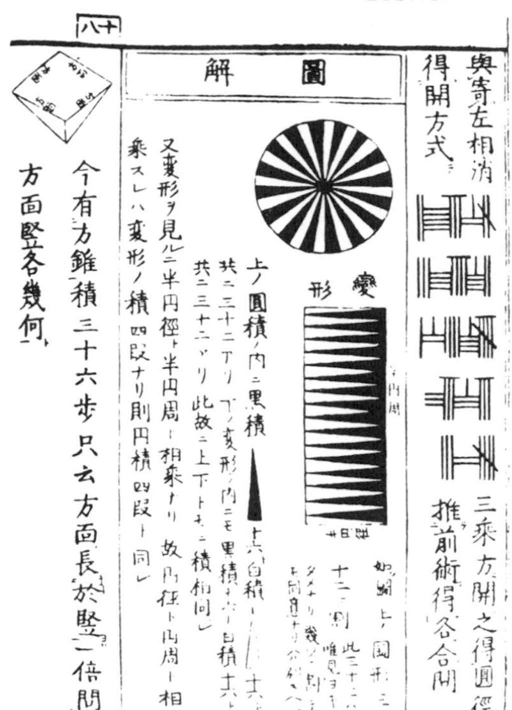

The rearrangement method used in a 17th century Japanese document.[4]

much more closely; and the area of the circle is still exactlly one half of the quasi-parallelogram (c).

On continuing this process by cutting up the original circle into a larger and larger number of segments, the side formed by the little arcs of the segments will become indistinguishable from a straight line, and the quasi-parallelogram will turn into a true parallelogram (a rectangle) with sides $2\pi r$ and r. Hence the area of the circle is half of this rectangle, or πr^2.

The same construction can be seen in the Japanese document above (1698). Leonardo da Vinci also used this method in the 16th century. He did not have much of a mathematical education, and in any case, he could use little else, for Europe in his day, debilitated by more than a millenium of Roman Empire and Roman Church, was on a mathematical level close to that achieved in ancient Mesopotamia. It seems probable, then, that this was the way in which ancient peoples found the area of the circle.

And that should be our last speculation. From now on, we can rely on recorded history.

2

THE BELT

Accurate reckoning — the entrance into the knowledge of all existing things and all obscure secrets.

AHMES THE SCRIBE
17th century B.C.

MAN is not the only animal that uses tools on his environment; so do chimpanzees and other apes (also, some birds). As long as man was a hunter, the differences between the naked ape and the hairy apes was not very radical. But roughly about 10,000 B.C., the naked ape learned to raise crops and to tame other animals, and thereby he achieved something truly revolutionary: Human communities could, on an average, produce so much more food above the subsistence minimum that they could free a part of their number for activities not directly related to the provision of food and shelter.

This Great Agricultural Revolution first took place where the geographical conditions were favorable: Not in the north, where the winters were long and severe, and the conditions for farming generally adverse; nor in the tropics, where food was plentiful, clothing unnecessary, shelter easily available, and therefore no drastic need for improvement; but in the intermediate belt, where conditions were sufficiently adverse to create pressures for change, yet not so adverse as to foil the attempts of farming and livestock raising.

This intermediate belt stretched from the Mediterranean to the Pacific. The Great Revolution first took place in the big river valleys of Mesopotamia; later the Belt stretched from Egypt through Persia and India to China. States developed. Specialists came into being. Soldiers. Priests. Administrators. Traders. Craftsmen. Educators.

And Mathematicians.

The hunters had neither time nor need for ratios, proportionalities or conic sections. The new society needed surveyors and builders, navigators and timekeepers (astronomers), accountants and stock-keepers, planners and tax collectors, and, yes, mumbo-jumbo men to impress and bamboozle the uneducated and oppressed. This was the fertile ground in which mathematics flourished; and it is therefore not surprising that the cradle of mathematics stood in this Belt.

SINCE Mesopotamia was the first region of the Belt where the agricultural revolution occurred and a new society took hold, one would expect Babylonian mathematics to be the first and most advanced. This was indeed the case; the older literature on the history of mathematics often saw the Egyptians as the founders of mathematics, but this was due to the fact that more and earlier Egyptian documents than any others used to be available. The research of the last few decades has changed this, and as a small sidelight, we find a better approximation for π in Mesopotamia than in Egypt.

One of the activities for which the new society freed some of its members was, unfortunately, organized warfare, and the various peoples inhabiting the region at different times, Sumerians, Babylonians, Assyrians, Chaldeans and others, warred against each other as well as against outsiders such as Hittites, Scythians, Medes and Persians. The city of Babylon was not at all times the center of this culture, but the mathematics coming from this region is simply lumped together as "Babylonian."

In 1936, a tablet was excavated some 200 miles from Babylon. Here one should interject that the Sumerians were first to make one of man's greatest inventions, namely, writing; through written communication, knowledge could be passed from one person to others, and from one generation to the next and future ones. They impressed their cuneiform (wedge-shaped) script on soft clay tablets with a stylus, and the tablets were then hardened in the sun. The mentioned tablet, whose translation was partially published only in 1950,[5] is devoted to various geometrical figures, and states that the ratio of the perimeter of a regular hexagon to the circumference of the circumscribed circle equals a number which in modern notation is given by $57/60 + 36/(60)^2$ (the Babylonians used the sexagesimal system, i.e., their base was 60 rather than 10).

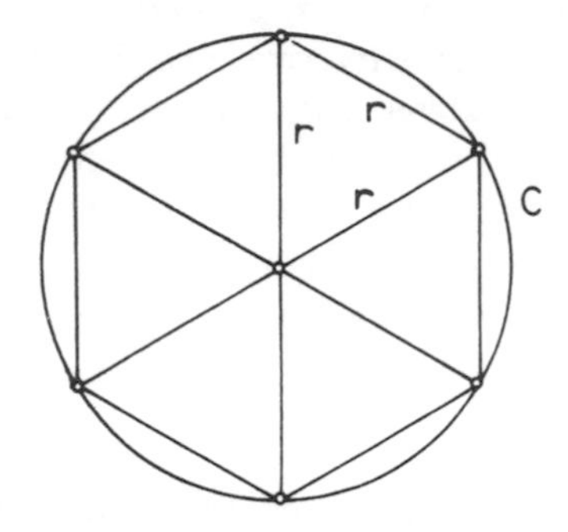

The Babylonian value of π.

The Babylonians knew, of course, that the perimeter of a hexagon is exactly equal to six times the radius of the circumscribed circle, in fact that was evidently the reason why they chose to divide the circle into 360 degrees (and we are still burdened with that figure to this day). The tablet, therefore, gives the ratio $6r/C$, where r is the radius, and C the circumference of the circumscribed circle. Using the definition $\pi = C/2r$, we thus have

$$\frac{3}{\pi} = \frac{57}{60} + \frac{36}{(60)^2}$$

which yields

$$\pi = 3\tfrac{1}{8} = 3.125, \tag{1}$$

i.e., the value the Babylonians must have used for π to arrive at the ratio given in the tablet. This is the lower limit of our little thought experiment on p. 15, and a slight underestimation of the true value of π.

MORE is known about Egyptian mathematics than about the mathematics of other ancient peoples of the pre-Hellenic period. Not because they had more of it, nor because more of their documents have been found, but because the key to the Egyptian hieroglyphs was discovered much earlier than for the other cultures. In 1799, the Napoleonic expedition to Egypt found a trilingual tablet at Rosetta near Alexandria, the so-called Rosetta stone. Its message was recorded in Greek, Demotic and Hieroglyphic. Since Greek was known, the code was cracked, and decipherment of the Egyptian hieroglyphs proceeded rapidly during the last century. The Babylonian tablets are more durable, and tens of thousands have been unearthed; the university libraries of Columbia and Yale, for example, have large collections. However, an analogous trilingual stone in Persian, Medean and Assyrian was not deciphered until about 100 years ago (with Persian known, and its writing system related to Babylonian), and as far as the history of mathematics is concerned, substantial progress in deciphering tablets in cuneiform script was not made until the 1930's. Even now, a large quantity of the available tablets await investigation.

The oldest Egyptian document relating to mathematics, and for that matter, the oldest mathematical document from anywhere, is a papyrus roll called the Rhind Papyrus or the Ahmes Papyrus. It was found at Thebes in a room of a ruined building and bought by a Scottish antiquary, Henry Rhind, in a Nile resort town in 1858. Four years later, it was purchased from his estate by the British Museum, where it is now, except for a few fragments which unexpectedly turned up in 1922 in a collection of medical papers in New York, and which are now in the Brooklyn Museum.

Histories like these make one wonder how many such priceless documents have been used up by the Arabs as toilet papyri. The sad story of how some of these papyri and tablets have survived millenia only to be rendered worthless by the excavators of our time is told by Neugebauer.[6]

The Ahmes Papyrus contains 84 problems and their solutions (but often no hint on how the solution was found). The papyrus is a copy of an earlier work and begins thus:[7]

Accurate reckoning. The entrance into the knowledge of all existing things and all obscure secrets. This book was copied in the year 33, in the 4th month of the inundation season, under the King of Upper and Lower Egypt "A-user-Re," endowed with life, in likeness to writings made of old in the time of the King of Upper and Lower Egypt Ne-mat'et-Re. It is the scribe Ahmes who copies this writing.

From this, Egyptologists tell us, we know that Ahmes copied the "book" in about 1650 B.C. The reference to Ne-mat'et-Re dates the original to between 2,000 and 1800 B.C., and it is possible that some of this knowledge may have been handed down from Imhotep, the man who supervised the building of the pyramids around 3,000 B.C.

A typical problem (no. 24) runs like this:

A heap and its 1/7 part become 19. What is the heap?

The worked solution accompanying the problem is this:

Assume 7.
(Then) *1* (heap is) *7*
(and) 1/7 (of the heap is) *1*
(making a) *Total* *8*
(But this is not the right answer, and therefore)
As many times as 8 must be multiplied to give 19, so many times 7 must be multiplied to give the required number.

The text then finds (by rather complicated arithmetic) the solution $7 \times {}^{19}/_{8} = 16\,{}^{5}/_{8}$.

Today we would solve the problem by the equation

$$x + x/7 = 19,$$

where x is the "heap." The Egyptian method, called *regula falsi*, i.e., assuming a (probably wrong) solution and then correcting it by proportionality, is no longer used in high school algebra. However, one version of *regula falsi*, called *scaling*, is still very advantageous for some electrical circuits (see figure below).

But the problem of interest here is no. 50. Here Ahmes assumes that the area of a circular field with a diameter of 9 units is the same as the area of a square with a side of 8 units. Using the formula for the area of a circle $A = \pi r^2$, this yields

$$\pi\,(9/2)^2 = 8^2,$$

and hence the Egyptian value of π was

$$\pi = 4 \times (8/9)^2 = 3.16049\ldots,$$

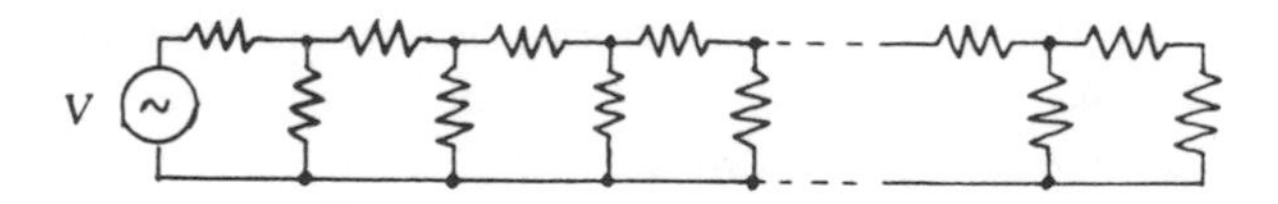

20th century version of Ahmes' method. The calculation of the currents due to a known voltage V in the network above becomes very cumbersome when the circuit is worked from left to right. So one assumes a (probably wrong) current of 1 ampere in the last branch and works the circuit from right to left (which is much easier); this ends up with the wrong voltage, which is then corrected and all currents are scaled in proportion. Just as was done by the Egyptians in 2,000 B.C.

The circuit above is also a mathematical curiosity for another reason. If all its elements are 1 ohm resistors, and the current in the last branch is 1 ampere, then the voltages across the resistors (from right to left) are Fibonacci numbers (1, 1, 2, 3, 5, 8, 13, 21, 34, 55, . . . , each new member being the sum of the last two).

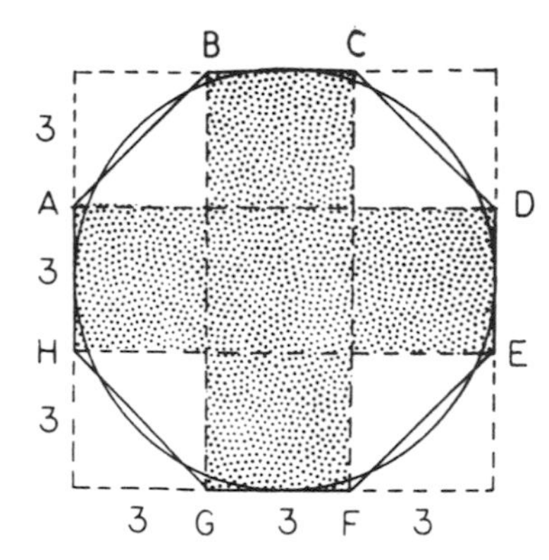

Egyptian method of calculating π.

a value only very slightly worse than the Mesopotamian value 3 1/8, and in contrast to the latter, an overestimation. The Egyptian value is much closer to 3 1/6 than to 3 1/7, suggesting that it was neither obtained nor checked by experimental measurement (which, as on p. 13, would have given a value between 3 1/7 and 3 1/8).

How did the Egyptians arrive at this weird number? Ahmes provides a hint in Problem 48. Here the relation between a circle and the circumscribed square is compared. Ahmes forms an (irregular) octagon by trisecting the sides of a square with length 9 units (see figure above) and cutting off the corner triangles as shown. The area of the octagon *ABCDEFGH* does not differ much from the area of the circle inscribed in the square, and equals the area of the five shaded squares of 9 square units each, plus the four triangles of 4½ square units each. This is a total of 63 square units, and this is close to 64 or 8^2. Thus the area of a circle with diameter 9 approximately equals 64 square units, i.e., the area of a square with side 8, which, as before, will lead to the value $\pi = 4 \times (8/9)^2$.

Ahmes has, of course, swindled twice: once in setting the area of the octagon equal to the area of the circle, and again in setting $63 \approx 64$. It is, however, noteworthy that these two approximations partially compensate for each other. Indeed, for a square of side a, the area of the octagon is $7(a/3)^2$, and this is to equal p^2, where p is the side of the square with equal area. If we now swindle only once by setting the area of the octagon equal to that of the circle, then

$$\tfrac{1}{4}\pi a^2 = p^2 = 7(a/3)^2,$$

and from the first and last expressions we find $\pi = 28/9 = 3\ 1/9$. This is a much worse approximation that Ahmes', confirming the time-honored rule that the last error in a calculation should cancel out all the preceding ones.

THE agricultural revolution in the Indian river valleys probably took place at roughly the same time as along the Nile valley and the valley of the Euphrates and Tigris (Mesopotamia). The apparent delay of Indian mathematics behind that of Babylon, Egypt and Greece may well be simply due to our ignorance of early Indian history. There is much indirect evidence of Hindu mathematics. Like everybody else in the Belt, they knew Pythagoras' Theorem long before Pythagoras was born, and there is evidence that Hindu astronomy was at a highly advanced level. Direct records, however, have been lost, and the first available documents are the Siddhantas or systems (of astronomy), published about 400 A.D., though the knowledge contained in them is of course much older.

One of the Siddhantas, published in 380 A.D., uses the value

$$\pi = 3\ 177/1250 = 3.1416$$

which differs little from the sexagesimal value

$$\pi = 3 + 8/60 + 30/(60)^2$$

used by the Greeks much earlier.

Early Hindu knowledge was summarized by Aryabhata in the *Aryabhatiya*, written in 499 A.D. This gives the solutions to many problems, but usually without a hint of how they were found. One statement is the following: [8]

Add 4 to 100, multiply by 8, and add 62,000. The result is approximately the circumference of a circle of which the diameter is 20,000.

This makes

$$\pi = 62{,}832 : 20{,}000 = 3.1416$$

as in the Siddhanta. The same value is also given by Bashkara (born 1114 A.D.), who calls the above value "exact," in contrast to the "inexact" value 3 1/7.

It is highly likely[9] that the Hindus arrived at the value above by the Archimedean method of polygons, to which we shall come in Chapter 6. If the length of the side of a regular polygon with n sides inscribed in a circle is $s(n)$, then the corresponding length for $2n$ sides is

$$s(2n) = \sqrt{2 - \sqrt{4 - s^2(n)}}$$

Starting (naturally) with a hexagon, progressive doubling leads to polygons of 12, 24, 48, 96, 192 and 384 sides. On setting the diameter of the circle equal to 100, the perimeter of the polygon with 384 sides is found to be the square root of 98694, whence

$$\pi \approx \sqrt{98694}/100 = 3.14156...,$$

which is the value given by Arayabatha.

The Hindu mathematician Brahmagupta (born 598 A.D.) uses the value

$$\pi \approx \sqrt{10} = 3.162277...$$

which is probably also based on Archimedean polygons. It has been suggested[10] that since the perimeters of polygons with 12, 24, 48 and 96 sides, inscribed in a circle with diameter 10, are given by the sequence

$$\sqrt{965}, \sqrt{981}, \sqrt{986} \text{ and } \sqrt{987},$$

the Hindus may have (incorrectly) assumed that on increasing the number of sides, the perimeter would ever more closely approach the value $\sqrt{1000}$, so that

$$\pi = \sqrt{1000}\ /10 = \sqrt{10}.$$

WHAT has been said about the ancient history of mathematics in India, and our ignorance of it, applies equally well to the Pacific end of the Belt, China. However, study of this subject has recently been facilitated by the impressive work of Needham.[11]

As in the West (see p. 14), the value $\pi \approx 3$ was used for several centuries; in 130 A.D., Hou Han Shu used $\pi = 3.1622$, which is close to $\pi = \sqrt{10}$. (The Chinese were singular among the ancient peoples in that they used the decimal system from the very beginning.) A document in 718 A.D. takes $\pi = 92/29 = 3.1724 \ldots$ Liu Hui (see figure on next page), in 264 A.D., used a variation of the Archimedean inscribed polygon; using a polygon of 192 sides, he found

$$3.141024 < \pi < 3.142704$$

and with a polygon of 3,072 sides, he found $\pi = 3.14159$.

18th century explanation of Liu Hui's method (264 A.D.) of calculating the approximate value of π.[12]

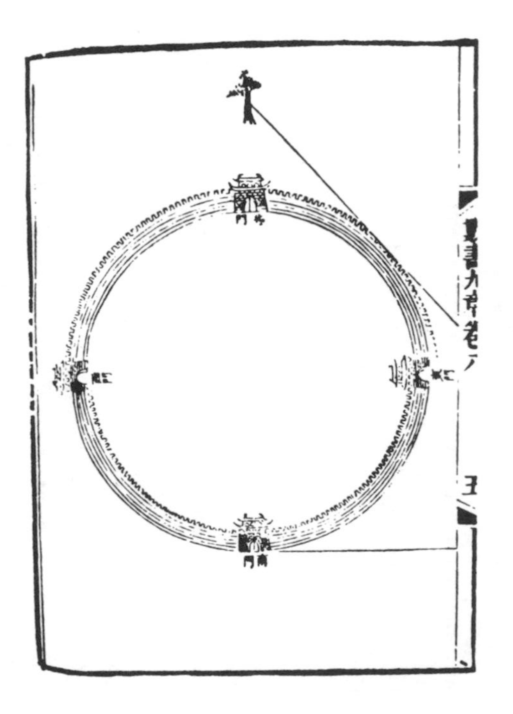

Determination of the diameter and circumference of a walled city from a distant observation point (1247 A.D.).[13]

In the 5th century, Tsu Chung-Chih and his son Tsu Keng-Chih found

$$3.1415926 < \pi < 3.1415927,$$

an accuracy that was not attained in Europe until the 16th century.

Not too much should be made of this, however. The number of decimal places to which π could be calculated was, from Archimedes onward, purely a matter of computational ability and perseverance. Some years ago, it was only a matter of computer programming know-how; and today it is, in principle, no more than a matter of dollars that one is willing to spend for computer time. The important point is that the Chinese, like Archimedes, had found a *method* that, in principle, enabled them to calculate π to any desired degree of accuracy.

Yet the high degree of accuracy that the Chinese attained is significant in demonstrating that they were far better equipped for numerical calculations than their western contemporaries. The reason was not that they used the decimal system; the decimal system by itself proves nothing except that nature was a poor mathematician in giving us ten fingers (instead of a number that has more integral factors, like twelve). As human language shows, everybody used base ten for numeration (or its multiple, 20, as in French and Danish).

But the Chinese discovered the equivalent of the digit zero. Like the Babylonians, they wrote numbers by digits multiplying powers of the base (10 in China, 60 in Mesopotamia), just like we do. But where the corresponding power of 10 was missing (102 has no tens), they left a space. The Hindus later used a circle for the digit zero (0), and this reached Europe via the Arabs and Moors only in the late Middle Ages (Italy) and the early Renaissance (Britain). An edict of 1259 A.D. forbade the bankers of Florence to use the infidel symbols, and the University of Padua in 1348 ordered that price lists of books should not be prepared in "ciphers," but in "plain" letters (i.e., Roman numerals).[14] But until the infidel digit 0 was imported, few men in Europe had mastered the art of multiplication and division, let alone the extraction of square roots which was needed to calculate π in the Archimedean way.

THE Belt, as we have argued, was the region conducive to the Great Agricultural Revolution that turned human communities from packs of hunters to societies with surplus productivity, freeing some of its members for activities other than provision of food. If this Belt stretched from the valley of the Nile to the Pacific, why should it not also cross America?

Indeed, it did. The agricultural revolution had taken place, at times comparable to the origins of the Afro-Asian Belt, in parts of Central and South America, where the impressive civilizations of the Aztecs, the Maya, the Chibcha, the Inca and others had grown up. A Maya ceremonial center in Guatemala, by carbon 14 tests, dates back to 1182 ($\pm$ 240) B.C,[15] and agriculture had of course begun much earlier; in fact, it was the American Indian who first domesticated two of our most important crops — corn (maize) and the potato.[16]

The most advanced of these cultures was that of the Maya, who established themselves in the Yucatan peninsula of Mexico, parts of Guatemala, and western Honduras. Judging by the Maya calendar, Maya astronomy must have been as good as that of early Egypt, for by the first century A.D. (from which time we have some dated documents), they had developed a remarkably accurate calendar, based on an ingenious intermeshing of the periods of the Sun, the Moon and the Great Star *noh ek* (Venus). The relationship between the lunar calendar and the day count was highly accurate: The error amounted to less than 5 minutes per year. The Julian calendar, which had been introduced in Rome in the preceding century, and which some countries of the Old World retained up to the 20th century, was in error by more than 11 minutes per year. From this alone it does not follow that the Maya calendar was more accurate than the Julian calendar (as some historians have concluded), but the point has some bearing, albeit very circumstantial, on the value that the Maya might have used for π, and we shall digress to examine the history of our calendar.

A calendar is a time keeping device. If it is to be of practical use beyond a prediction of when to observe a religious holiday, it must be matched to the seasons of the year (i.e., to the earth's orbit round the sun) very accurately: If we lose an hour or two per calendar year with respect to the time it takes the earth to return to the same point on its orbit, this does not seem a significant error; yet the difference will accumulate as the years go by, and eventually one will find that at 12 noon on a summer day (by this calendar) it is not only dark, but it is snowing as well.

Astronomy for calendar making was therefore one of the earliest activities involving mathematics in all ancient societies. In Egypt and Babylon, as well as in Maya society, the priests had a monopoly of learning. In all three societies the priests were the astronomers, calendar makers and time keepers.

The Babylonians first took a year equal to 360 days (presumably because of their sexagesimal system and their 360° circle), but later they corrected this by 5 additional days. This value was also adopted by the Egyptians, and it was to be the value used by the Maya. To correct this value to a fraction of a day, it was necessary to observe the movement of the other stars on the celestial sphere. The Maya watched the planet Venus, the Egyptians the fixed star Sirius. Owing to the precession and nutation of the earth's axis, Sirius is not really fixed on the celestial sphere (tied to the coordinates of the terrestrial observer), but has, as viewed from the earth, a motion of its own. The Egyptians found that Sirius moved exactly one day ahead every 4 years, and this enabled them to determine the length of one year as 365¼ days. One of the Greco-Egyptian rulers of Egypt in the Hellenic age, Ptolemy III Euergetes, a mathematician whom we shall meet again, issued the following decree in 238 B.C.:

> Since the Star [Sirius] advances one day every four years, and in order that the holidays celebrated in the summer shall not fall into winter, as has been and will be the case if the year continues to have 360 and 5 additional days, it is hereby decreed that henceforth every four years there shall be celebrated the holidays of the Gods of Euergetes after the 5 additional days and before the new year, so that everyone might know that the former shortcomings in reconing the seasons of the year have henceforth been truly corrected by King Euergetes. [17]

But by this time Egyptian priesthood had become more interested in religious mumbo jumbo than in science, and they sabotaged the enforcement of Ptolemy's decree. Ironically, it fell to the Roman adventurer, warlord and vandal Gaius Julius Caesar to bring about the adoption of the leap year. Alexandria, in the 3rd to 1st centuries B.C., was the intellectual center of the ancient world, the like of which had not been seen before and was not to be seen again until the rise of Cambridge and the Sorbonne. In 47 B.C., Caesar's hordes ransacked the city and burned its libraries, and Caesar, during whatever time he had left between his romance with Cleopatra and contemplating further conquests, managed to get acquainted with the astounding achievements of Alexandrian astronomers. He took one of them, Sosigenes, back to Rome and inaugurated the Julian calendar as of January 1, 45 B.C. The new calendar introduced Ptolemy's leap year, giving a year an average duration of 365.25

days. But the true duration of the earth's orbit about the sun is about 365.2422 days. (Actually, the earth is subject to all kinds of perturbations, and today we do not use its orbit as a standard any more. "Atomic clocks" run much more accurately than the earth. When such an atomic clock appears to be x seconds late at the end of a tropical year, we say that the earth has completed its orbit x seconds early.)

The difference between the two values, a little over 11 minutes per year, accumulated over the centuries and once more threatened to put the date out of date. In 1582, Pope Gregory XIII decreed that the extra day of a leap year was to be omitted in years that are divisible by 100, unless also divisible by 400 (i.e., omitted in 1800, 1900, but not in 2000). This is the calendar we are using now. It was soon adopted by the Catholic countries, but the others thought it "better to disagree with the Sun than to agree with the Pope," and Britain, for example, did not adopt it until 1752. By that time the British had slipped behind 11 days, and when they were simply omitted to catch up with the rest of Europe, many Britons were outraged, accusing the government of conspiring to shorten their lives by 11 days and to rob them of the interest on their bank accounts. Russia held on to the Julian calendar until the October Revolution, which took place in November (1917).

Getting further and further away from the Maya, we may note that our present calendar is far from satisfactory. Not because it is still off by 2 seconds per year, for these accumulate to a full day only once in 3,300 years; but because our present calendar is highly irregular: A date falls on a different day of the week every year, and the months (even the quarters) have different lengths, which complicates, among other things, accounting and other business administration. In 1923 the League of Nations established a Committee for Calendar Reform, which, predictably, achieved nothing, and the same result has hitherto been achieved by the United Nations. Of the hundreds of submitted proposals, UNESCO in 1954 recommended the so-called "World Calendar" for consideration by the UN General Assembly. The proposal does away with both of the mentioned disadvantages of our present calendar, yet does not change it so radically as to cause widespread confusion. Most governments agreed to the proposal in principle, but some (including the US) considered it "premature" and the matter is still being "considered." The UN, a grotesque assembly of propaganda-bent hacks, has found itself unable to condemn international terrorism by criminals, much less to reform the calendar.

Maya glyph denoting position of month in half-year period.

It is against this background of difficulties with calendar making that the achievements of the Maya must be viewed. True, their calendar year, like that of the Babylonians and Egyptians, amounted to 365 days, so that their religious festivals drifted with respect to the natural seasons. Like the Babylonians and Egyptians, they must have known that they were gaining one day every four years, but like the Egyptians, they evidently preferred keeping their religious holidays intact to using the calendar as a time keeper for sowing, harvesting and other activities geared to the periodicity of nature.

If we do not require a calendar to be geared to a tropical year (earth's orbit), but only that it be geared to *some* part of the celestial clock, then the Maya calendar was more accurate than the Julian calendar, more accurate than the Egyptian (solar) calendar, and more accurate than the Babylonian (solar-lunar) calendar; it intermeshed the "gear wheels" of Sun, Moon and Venus, and was based on a more accurate "gear ratio" than the other calendars, repeating itself only once in 52 years.

It is unthinkable that a people as advanced in astronomy should not have come across the problem of calculating the circle ratio. If their astronomy was as advanced as that of the Egyptians, is it not reasonable to assume that the Maya value of π was as good as that of the Egyptians?

It is not. It is reasonable to assume that it was far better. For the Maya were incomparably better equipped for numerical calculations than the Egyptians were; they had discovered the zero digit and the positional notation that had escaped the genius of Archimedes, and that held up European arithmetic for a thousand years after the Maya were familiar with it. They used the vigesimal system (base 20), the digits from 1 to 19 being formed by combinations of ones (dots) and fives (bars), as shown in the figure below.

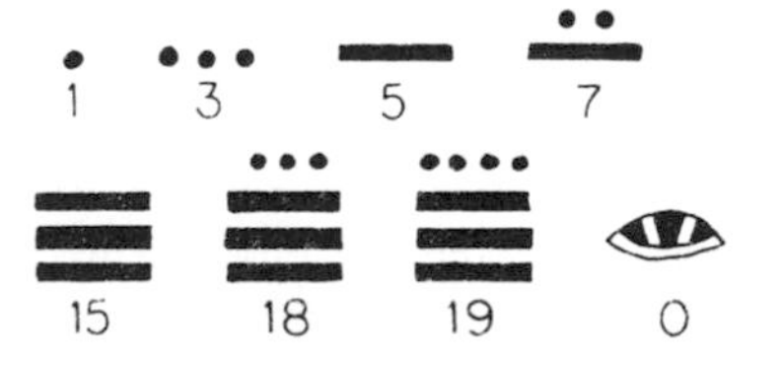

Maya digits

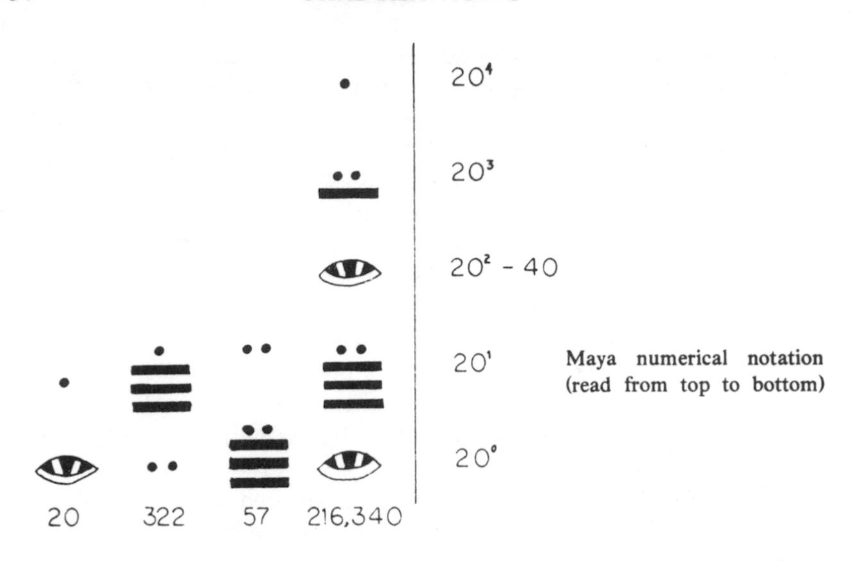

Maya numerical notation (read from top to bottom)

Any positionally expressed number *abcd.ef* using base x denotes the number

$$ax^3 + bx^2 + cx^1 + dx^0 + ex^{-1} + fx^{-2}.$$

The Maya notation fits this pattern with $x = 20$ and the digits shown on p. 33, with an exception for the second-order digit. In a pure vigesimal system, this would multiply $20^2 = 400$, but for reasons connected with their calendar, the Maya used this position to multiply the number 18×20 or $20^2 - 40$. Also, it is not known whether they progressed beyond a "vigesimal point" to negative powers of 20, i.e. to vigesimal fractions. Examples of the Maya notation are given in the figure above. However this may be, it is clear that with a positional notation closely resembling our own of today, the Maya could out-calculate the Egyptians, the Babylonians, the Greeks, and all Europeans up to the Renaissance.

The Chinese, who had also discovered the digit zero and the positional notation utilizing it, had found the value of π to 8 significant figures a thousand years before any European. The Maya value might have been close to that order.

But we can only guess. The Maya civilization went the way of other civilizations; after it peaked, it decayed. About the 7th century A.D., the Maya started to desert their temple cities, and within a century or two these splendid cities were abandoned, no one knows why. Civil strife set in, and later they were conquered by the Aztecs. By the time the Spaniards arrived, they were far down from their classical age.

There were some records, of course. The Maya wrote books on long strips of bark or parchment, folded like a screen. How many of these dealt with their mathematics, geometry and the circle ratio, we shall never know. In the 1560's, Diego de Landa, Bishop of Yucatan, burned the literature of the Maya on the grounds that "they contained nothing in which there were not to be seen superstition and lies of the devil."[18] What remained was burned by the natives who had been converted to the Bishop's religion of love and tolerance.

Today the American Indians, in their quest for identity and self-respect, complain that even the name of their people was given them "by some honkey who landed here by mistake."

They might add that the Red Man had made the great discovery of positional notation employing the digit zero a thousand years before the Palefaces; at a time when Spain was a colony of a benighted empire, and when the ancestors of the Anglo-Saxons were illiterate hunters in the virgin forests of continental Europe, no one knows exactly where.

3

The Early Greeks

O King, for traveling over the country, there are royal roads and roads for common citizens; but in geometry there is one road for all.

MENAECHMUS (4th century B.C.), when his pupil Alexander the Great asked for a shortcut to geometry.

IN following the Belt along its length, we have lost the thread of time; and we now return to the Eastern Mediterranean, where the history of mankind went through a few inspiring centuries associated with the ancient Greeks. The Golden Age of this time was the age of the University of Alexandria; but just as the Alexandrian sun continued to glimmer some centuries after the Romans had sacked its places of learning and burned its libraries, so there was already some light before the city and its University were founded.

This was a time when the Greeks first held their own against the then all-mighty Persian Empire at Marathon (490 B.C.) and then defeated the Persians at Plataea (479 B.C.). This was also the time when democratic government developed; a slaveowners' democracy, yes, but a democracy. The next 150 years saw the confrontation of Athens and Sparta, the thinkers against the thugs. The thugs always win, but the thinkers always outlast them.

In mathematics, which is but a mirror of the society in which it thrives or suffers, the pre-Athenian period is one of colorful men and important discoveries. Sparta, like most militaristic states before and after it, produced nothing. Athens, and the allied Ionians, produced a number of works by philosophers and mathematicians; some good, some controversial, some grossly overrated.

As far as the history of π is concerned, there were four men of this period who had some bearing on the problem: Anaxagoras, Antiphon, Hippocrates and Hippias.

Anaxagoras of Clazomenae (500-428 B.C.) found in Athens a bold new spirit of free enquiry and enthusiastically joined in with a truly scientific spirit, not shying away from experiment nor from popularizing science (both in diametrical opposition to the snobbism of the later Greek philosophers). He taught that the annual swelling of the Nile was due to the melting of mountain snows near the upper part of the river; that the Moon received its light solely from the Sun; that eclipses of the Moon or the Sun were caused by the interposition of the Earth or the Moon; that the Sun was a red-hot stone bigger than the entire Peloponnese; and other (less successful) theories. But in his theory of the Sun, which denied that the Sun was a deity, he had gone too far, and for some time he was imprisoned in Athens for impiety.

While in jail, he attempted to "square the circle," a problem intimately connected with π, which requires the construction of a square equal in area to a given circle, but which we shall examine in more detail in the next chapter. It is apparently the first mention (by Plutarch) of this famous problem that fascinated men until 1882, when the proof that it could not be solved (by Greek geometry) was finally furnished; and even now it continues to fascinate amateur circle squarers who cannot be dissuaded by mathematical proof.

Another Greek of this period (late 5th century B.C.) associated with squaring the circle is the Sophist philosopher Antiphon, who enunciated the "principle of exhaustion," which was to have a profound influence on mathematicians in their quest for the value of π until the invention of the calculus in the late 17th century. The exhaustion principle states the following: If a square is inscribed in a circle, and further regular polygons are inscribed with double the number of sides at each step (octagon, 16-sided polygon, etc.), until the circle is *exhausted*, then eventually a polygon will be reached whose sides are so short that it will coincide with the circle. This led Antiphon to believe that the circle could be squared by Greek geometry; for any polygon can be squared, and since the "eventual" polygon is equivalent to a circle, Antiphon judged that a circle could be squared also.

Antiphon's principle is almost correct; it becomes completely correct in Euclid's more cautious formulation, according to which the difference between the remaining area and that of the circle can be

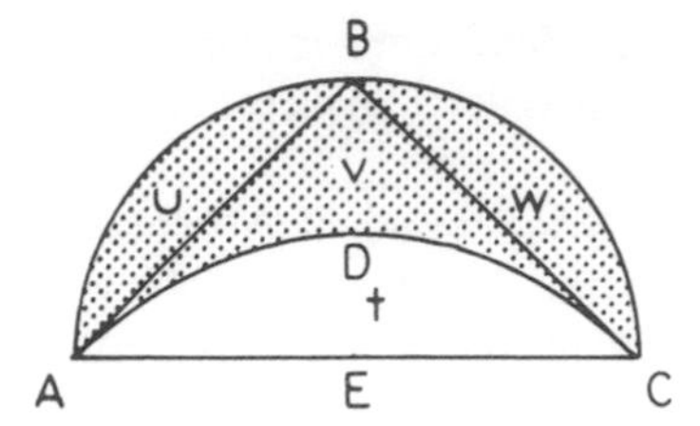

Hippocrates lune. The lower case letters stand for areas.

made smaller than any preassigned value, no matter how small (but not zero).

Another man associated with the area of circles during this period was Hippocrates of Chios, who should not be confused with his better known contemporary, the physician Hippocrates of Cos. Hippocrates of Chios was a merchant who came to Athens in 430 B.C. According to some, he lost his money through fraud in Byzantium, according to others he was robbed by pirates; in any case, he thereafter turned to the study of geometry, which is in itself remarkable, for in later years the road between business and science was more often traveled in the opposite direction.

Hippocrates was evidently the first to find the *exact* area of a figure bounded by curves, in this case circular arcs, and thus became the first man to make a precise statement on curvilinear mensuration. The area that Hippocrates squared (or triangled, but to square a triangle is trivial) was that of a lune (see figure above) bounded by the semicircular arc *ABC* circumscribed about the right-angled, isosceles triangle *ABC*, and the circular arc *ADC* such that the segment *ADEC* is similar to the segment *AB*. (That is, the radius of *ADC* is to *AC* as the radius *EA* is to *AB*.)

Hippocrates found that the area of this lune is exactly equal to the triangle *ABC*.

Proof: Since the segments above the three sides are similar, their areas are to each other as the squares of their bases; since by Pythagoras' Thorem $AB^2 + BC^2 = AC^2$, the areas of the corresponding segments are related by

$$t = u + w. \tag{1}$$

The area of the lune is

$$l = u + v + w,$$

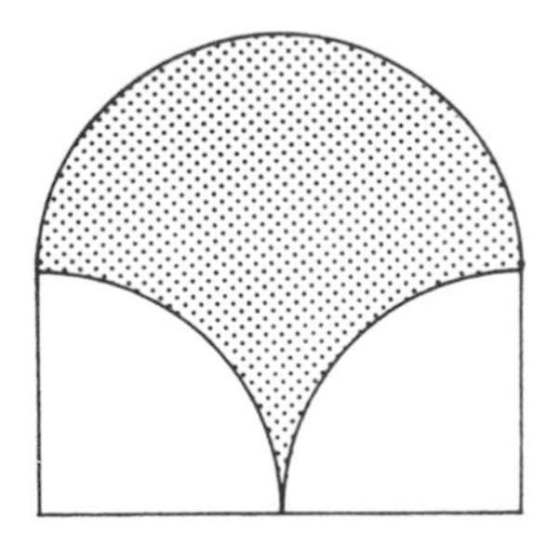

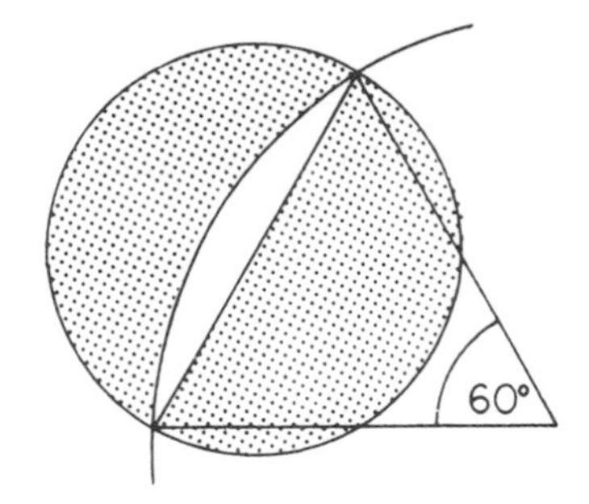

Examples of figures that are exactly squarable by Euclidean geometry. The shaded areas are reducable to Hippocrates lunes, and hence are squarable.

or on substituting (1),

$$l = v + t,$$

i.e., it is equal to the area of the triangle.

This is probably not the proof given by Hippocrates, whose book on geometry has been lost. The proof of this theorem given by Eudoxos and taken over by Euclid more than a century later is a *reductio ad absurdum* (reduction to the absurd), a proof by contradiction or indirect proof. It is possible that Hippocrates was the first to use this method of proof, whose essence is this: If we want to prove that a statement is true, we first assume that it is not true and use this assumption to achieve an absurd result (or a result that contradicts the assumption); since the result is absurd, the premise must have been false; if a statement is either true or false (*tertium non datur*, there is no third possibility), the statement must be true.

For example, the fact that the number of primes (numbers divisible only by themselves or one) is infinite is proved as follows. Assume that the number of primes is finite. Then there must exist the largest prime, call it p. Now the number

$$p! = p(p-1)(p-2)(p-3)\ldots 2 \cdot 1$$

is divisible by all integers up to and including p. Hence the number $p! + 1$ is divisible by none of them and is therefore a prime. But the prime $p! + 1$ is obviously greater than p, which contradicts the assumption that p is the largest prime. The assumption is therefore false, and the number of primes is infinite.

Hippocrates' discovery can easily be generalized to many other curvilinear figures, two of which are shown above. The one on the right, in particular, must have raised high hopes of squaring the circle (or the semicircle, which is just as good). The snag, as we know today, is that some, but not all, lunes can be squared. In the top right figure, for example, the sum of the lune and the semicircle can be squared, but they cannot be squared individually — at least not by the Greek rules of the game (which will be discussed in the next chapter).

Very many other figures composed entirely or partly of Hippocrates lunes can be constructed. Leonardo da Vinci (1452-1519) was particularly fascinated by Hippocrates lunes, and he constructed 176 such figures on a single page of his manuscripts, and more elsewhere.[19]

It is ironic that the work of Antiphon and Hippocrates should have been brought to our attention by one whose teachings held up the progress of science for close to 2,000 years, the Greek philosopher Aristotle (384-322 B.C.). He judges both Antiphon's principle of exhaustion and Hippocrates' quadratures to be false, considering the refutation of Antiphon's principle beneath the notice of geometers, and never, of course, achieving a disproof of Hippocrates' quadrature of the lune, either. Aristotle, we are invaribly told, was "antiquity's most brilliant intellect," and the explanation of this weird assertion, I believe, is best summarized in Anatole France's words: The books that everybody admires are the books that nobody reads. But on taking the trouble to delve in Aristotle's writings, a somewhat different picture emerges (see also Chapter 6). His ignorance of mathematics and physics, *compared to the Greeks of his time*, far surpasses the ignorance exhibited by this tireless and tiresome writer in the many subjects that he felt himself called upon to discuss.

THE fourth man of interest in this early Greek period is Hippias of Elis, who came to Athens in the second half of the 5th century B.C., and who was the first man on record to define a curve beyond the straight line and circle. It is perhaps ironic that the next curve on the list should be a transcendental one, skipping infinitely many algebraic curves, but the Greeks did not yet know about degrees of a curve, and so they ate fruit from all orchards. The curve that Hippias discovered was called a trisectrix, because it could be used to trisect an angle (the second problem that fascinated antiquity, the third being the doubling of the volume of a cube), or a quadratrix, because it could be used to square the circle, and with the mere use of compasses and straightedge, too. There has to be a snag, of course, and the snag was that the construction violated another rule of the Greek game, but this point is deferred to the next chapter.

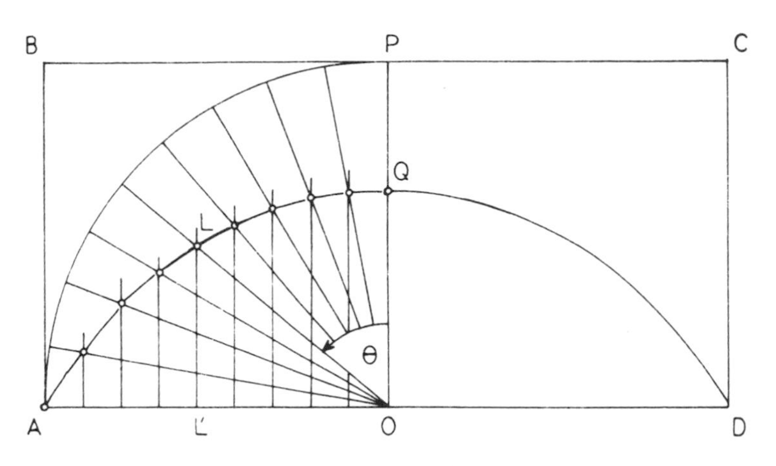

Hippias' quadratrix

Hippias' quadratrix was defined as follows. Let the straight segment *AB* move uniformly from the indicated position until it coincides with *CD*, and let the segment *AO* rotate uniformly clockwise about the point *O* from the indicated position through *OP* until it coincides with *OD*, in the same time as *AB* moves to *CD*. The curve traced by the intersection (*L*) of the two segments during this motion is Hippias' quadratrix.

Using only compasses and straightedge, and discarding the concepts of motion and time, we can also construct this curve by dividing the angle *AOP* into 2^n parts, and the segment *AO* into the same number of parts, where *n* is arbitrarily large. Any intersection of two corresponding segments is then a point on the quadratrix.

We do not know whether Hippias realized that by means of his curve the circle could be squared; perhaps he realized it, but could not prove it. The proof was later given by Pappus (late 3rd century A.D.), who evidently had it from Dinostratus (born about 350 B.C., brother of Menaechmus, who is quoted at the beginning of this chapter). His proof by elementary geometry is a *reductio ad absurdum*, but the modern road is shorter:

Let $OL = \rho$ and $OA = r$; from the definition of the quadratrix, and expressing the angle of rotation in radian measure, we have

$$\frac{AL'}{\pi/2 - \theta} = \text{const} = \frac{2r}{\pi}$$

and since $AL' = r - \rho \sin\theta$, we obtain

$$2r = \pi\rho(\sin\theta)/\theta. \tag{1}$$

For $\theta \to 0$, we have $(\sin\theta)/\theta \to 1$, $\rho \to OQ$; also $2r = AD$, so that (1) yields

$$AD : OQ = \pi, \tag{2}$$

and we have a geometrical construction for the circle ratio π.

To rectify the circle, we need a segment of length

$$u = 2\pi r = 2(AD : OQ) \times OA = (AD \times AD) : OQ$$

or

$$u : AD = AD : OQ,$$

so that u can be found by an easy construction of proportional quantites (similar triangles).

A rectangle with sides $u/2$ and r then has area

$$\tfrac{1}{2}ur = \pi r^2,$$

i.e., the same area as a circle with radius r, and since a rectangle is easily squared by an elementary construction, the circle is squared by the use of compasses and straightedge alone. Nevertheless, as we shall see, this construction failed to qualify under the implicit rules of Greek geometry.

HIPPIAS and Dinostratus only scratched the surface of this curve, for analytical geometry was still some 2,000 years away. But we might as well finish the job for them.

Let us regard (1) as the equation of the quadratrix in polar coordinates through

$$\rho = \sqrt{(x^2 + y^2)},$$
$$\phi = \arctan(y/x);$$

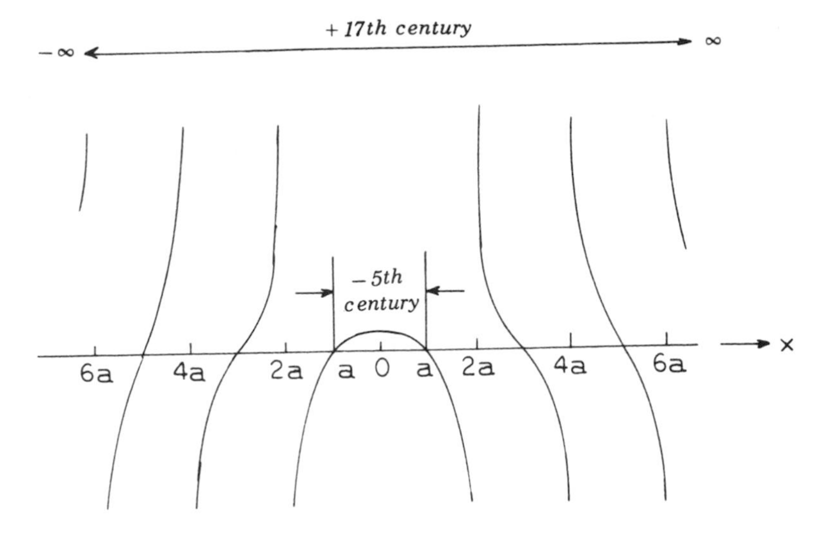

Hippias' quadratrix derived by analytical geometry. Hippias knew only the part in the interval $-a < x < a$, i.e., the one shown in the preceding figure.

After some manipulation, we have

$$y = x \cot(\pi x / 2a), \tag{3}$$

and the figure above shows how the quadratrix has bloomed. Hippias and Dinostratus only saw the tip of the iceberg for $-r < x < r$.

The circle can also be squared by the use of the Archimedean spiral, as we shall see.

The essence of Hippias' trick was the fact that the limit of $(\sin\theta)/\theta$ for $\theta \to 0$ is unity. But there are other curves exploiting this limit, and these can also be used as quadratrices. Three more appeared in the 17th century after Descartes' discovery of analytical geometry:

Tschirnhausen's quadratrix,

$$y = \sin(\pi x / 2r),$$

Ozam's quadratrix

$$y = 2r \sin^2(x/2r),$$

and the cochleoid, which in polar coordinates is given by

$$\rho = (r \sin\phi)/\phi.$$

But the problem has lost its fascination, and these curves have, for most people, faded into oblivion. However, when viewed as functions rather than curves, they are still making a living in electromagnetics, spectral analysis and information theory, occasionally moonlighting in statistics.

4

EUCLID

King Edward's new policy of peace was very successful and culminated in the Great War to End War. [It was followed by] the Peace to End Peace.

W.C. Sellar and R.J. Yeatman,
1066 and All That [20]

SOMETHING in the vein of the quotation above might also be said of Alexander the Great. He certainly had no policy of peace, but he, too, demonstrated the French saying that "the more this changes, the more it is the same thing." At age 32 he had conquered the world, at least the world known to antiquity, everything there was to conquer between Greece and India, and some more into the bargain. At age 33 he was dead, and so was his great empire: it broke up into a heap of little empires, each of which wanted to be the Great Empire and therefore fought all the other little empires.

During his expedition to Egypt in 332-331 B.C., Alexander had founded the city of Alexandria. After his death in 323 B.C., his generals fought each other over who was to get his hands on what, but by 306 B.C. control over Egypt had firmly been established by one of them, Ptolemy I; he was succeeded by his son Ptolemy II, who in 276 B.C. married his own sister Arsinoe. To marry one's own sister was a Pharaonic practice (though the Ptolemies were Greeks); but Ptolemy did something much more significant. He founded the Museum and Library of Alexandria. [21]

For the Library, Ptolemy acquired the most valuable manuscripts, had translations made of them, had purchasing agents scour the Mediterranean for valued books, and even compelled travelers arriving in Egypt to give up any books in their possession; they were copied by scribes in the Library, the original was retained, and the copy given to its owner. His son Ptolemy III (the one who decreed leap years, see p. 31) was even more avid: He borrowed the original copies of the works of famous Greek playwrights from the Athenians, had them copied, sent back the copies and cheerfully forfeited the deposit he had paid as bond for the return of the originals.[22] Before the arrival of Caesar's thugs, the Library had close to three quarters of a million books (that is, rolls; a standard roll was 15 to 20 feet in length and contained the equivalent of ten to twenty thousand words of modern English text[22]).

Even more impressive was the Museum. This was a school or institute — in effect, a university. Ptolemy engaged the most celebrated scholars of his time to teach at this university, and soon it became the scientific capital of the world. Few were the learned men of later antiquity who had not studied at Alexandria; they were taught by the finest scientists that the contemporary world could muster.

Mathematics flourished at Alexandria. Erastosthenes (273-192 B.C.), chief librarian, calculated the circumference of the earth to within 5% of the correct value by observing the difference in zenith distance of the sun at two places separated by a known distance (Alexandria and Syene, on approximately the same meridian); in posession of ever more accurate trigonometrical tables (obtained by working down from 90° by use of the half-angle formulas and then filling in the gaps by the addition theorems), he calculated the distance to the moon and to the sun. The method was correct, and although his imperfect measuring instruments yielded a large error for the moon, the ditance to the sun, as near as we can ascertain the length of his unit, the stadium, agrees with what we know today, including measurements by radar. And this was done at a time when the philosopher Epicurus in Athens taught that the sun was two feet in diameter!

The academic community at Alexandria was cosmopolitan, mainly Greek, Egyptian and Jewish. So was the city surrounding it. It was referred to as "Alexandria *near* Egypt," for it was not considered a part of Egypt proper. Indeed, it was not part of anything, evolving a culture of its own. Its citizens had a reputation of being lively, quick-witted, irreverent, and worthless as soldiers. The first three Ptolemies were unusually enlightened, but later decadence set in, and the line ended with the Quisling Queen Cleopatra. At times the Alexandrians would riot against a king they disliked and some of the

Preclarissimus liber elementorum Euclidis perspi/
cacissimi: in artem Geometrie incipit quã foelicissime:
Punctus est cuius ps nõ est. ¶ Linea est
lõgitudo sine latitudine cuiꝰ quidẽ ex/
tremitates st duo pũcta. ¶ Linea recta
ẽ ab vno pũcto ad aliũ breuissima extẽ/
sio i extremitates suas vtrũqʒ eoꝝ reci
piens. ¶ Superficies ẽ q̃ lõgitudinẽ ⁊ lati
tudinẽ tm̃ hʒ: cuiꝰ termi quidẽ sũt linee.
¶ Superficies plana ẽ ab vna linea ad a/
liã extẽsio i extremitates suas recipiẽs
¶ Angulus planus ẽ duarũ linearũ al/
ternus ꝯtactus: quaꝝ expãsio ẽ sup sup/
ficiẽ applicatioqʒ nõ directa. ¶ Quãdo aũt angulum ꝯtinet due
linee recte rectilineꝰ angulus noiat. ¶ Qñ recta linea sup rectã
steterit duoqʒ anguli vtrobiqʒ fuerĩt eq̃les: eoꝝ vterqʒ rectꝰ erit
¶ Lineaqʒ linee supstãs ei cui supstat ppendicularis vocat. ¶ An
gulus võ qui recto maior ẽ obtusus dicit. ¶ Angulꝰ võ minor re
cto acutꝰ appellat. ¶ Terminꝰ ẽ qd vniuscuiusqʒ finis ẽ. ¶ Figura
ẽ q̃ tmino vl terminis ꝯtinet. ¶ Circulꝰ ẽ figura plana vna q̃dem li/

De principijs p se notis: ⁊ pmo de diffini/
tionibus earundem.

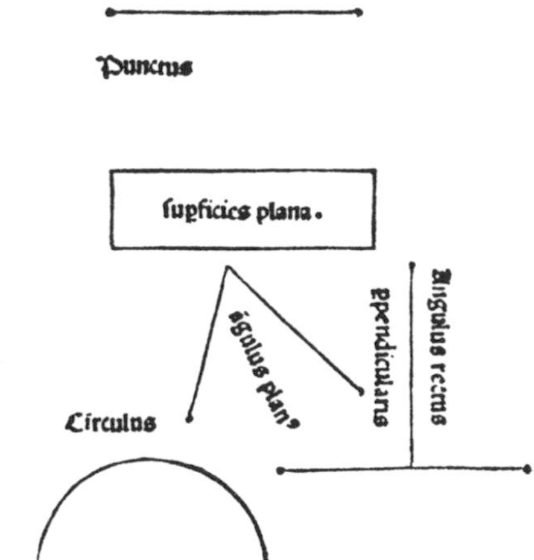

From a Latin edition of Euclid's *Elements* (1482). [82]

later kings would order a massacre. That, they must have laughed, would cure them. They are still laughing.

AMONG the scholars whom Ptolemy brought to Alexandria was Euclid, a man whose place and date of birth are unknown, so today he is simply called "Euclid of Alexandria." Euclid was, among other things, a publisher's dream. His *Elements* (of geometry) are the all-time bestseller of all textbooks ever written. More than a thousand editions have been published only since the invention of the letter-press in the 15th century, and the textbook, now in (roughly) its 2,250th year, is still going strong (especially considering that practically all school geometries are merely rehashes of Euclid's text).

The major part of the *Elements* was probably known before Euclid. But the importance of Euclid's work is not in what the theorems, as such, say. The great significance of the *Elements* lies in their method. The *Elements* are the first grandiose building of mathematical architecture. There are five foundation stones, or postulates, which (Euclid believed) are so simple and obvious that everyone can accept them. Onto these foundation stones Euclid lays brick after brick with iron logic, making sure that each new brick rests firmly supported by one previously laid, with not the tiniest gap a microbe can walk through, until the whole cathedral stands as firmly anchored as its foundations.

Euclid is not the father of geometry; he is the father of mathematical rigor.

The *Elements* consist of twelve books, and the elementary geometry taught in the world's high schools corresponds, in one form or another, to the first four. Alas, the reason why it should be taught is very often grossly misunderstood. Except for a few elementary theorems, Euclidean geometry is of little use for modern science and engineering: One can usually go much further and faster by trigonometry and analytical geometry. The real signifiance of Euclidean geometry lies in the superb training it gives for logical thinking. A proof must not contain anything that is ultimately based on what we want to prove, or the proof is circular and invalid. From "Every angle in a semicircle is a right angle" it does not follow that "The apex of a right angle subtended by the diameter lies on the circumference;" it so happens that the second statement is true, but it has to be proved. Euclidean geometry teaches the difference between *if* and *if and only if*, and the difference between *one* and *one and only one.* If somebody teaches you how to bisect an angle by the use of compasses without insisting on the proof that the construction is correct, he has thrown out the golden kernel and handed you the dead garbage. Who needs to bisect an angle, anyway? The few who do find a protractor faster and more accurate than using compasses three times over. But this is not the point; what matters is that the constructed line is the bisector and the only bisector.

And this brings us to the much misunderstood problem of squaring the circle, i.e., constructing a square whose area equals that of a given circle. There never was an Olympic Committee that laid down the rules of the game and the conditions under which the circle was to be squared. Yet from the way Euclid's cathedral is built, we know exactly what the Greeks had in mind. The problem was this:

(1) Square the circle,

(2) using straightedge and compasses only,

(3) in a finite number of steps.

It is evident why Hippias' and Dinostratus' method of squaring the circle was not accepted as "clean" by the Greeks, even though Dinostratus proved that it did square the circle. Not because it violated the second condition; it did not. Nor (as some write) because the quadratrix was a curve other than a straight line or a circle: Euclid himself wrote a book on conic sections (lost in the Roman catastrophe), and these were generally not circles, either. But referring to the figure on p. 41, it is easily seen that the point Q can only be found by placing a French curve along the points L, which violates rule 2, or constructing infinitely many points L by straightedge and compasses. Even assuming that the Greeks would have intuitively accepted the equivalent of the modern concept of continuity (and they accepted nothing but a handful of postulates), they meticulously steered clear of anything involving the infinite, for they were still puzzled by Zeno's paradoxes (about 500 B.C.), the best known of which is the one of Achilles and the tortoise: Achilles races the tortoise, giving it a headstart of 10 feet. While Achilles covers these 10 feet, the tortoise covers 1 foot; Achilles covers the remaining foot, but the elusive tortoise has covered another 1/10 of a foot, and so on; Achilles can never catch up with the tortoise. The stumbling block here was that the Greeks could not conceive of an infinite sum adding up to a finite number. In our own age, school children absorb the fact that this is so at the tender age when they first learn about decimal fractions:

$$10/9 = 1.1111\ldots = 1 + 1/10 + 1/100 + 1/1000 + 1/10000 + \ldots$$

As for the condition that only compasses and straightedge may be used in the construction, it has puzzled many, including some who ought to know better. It has been attributed to Greek esthetics and to the influence of Plato and Aristotle (from whom the Alexandrian scientists were mercifully removed by an ocean far larger than the Mediterranean); and other reasons have been suggested to explain the "Greek obsession with straightedge and compasses."

Nonsense. The only things the Greek mathematicians were obsessed with were truth and logic. Straightedge and compasses came into the problem only incidentally and as a secondary consequence, and esthetics had nothing to do with it at all. Just as Euclid built his cathedral on five foundation stones whose simplicity made them obvious, so one can go in the opposite direction and prove a statement by demonstrating that it was entirely supported by the stones of Euclid's building; from hereon downward it had already been proved that the building stones lie as firmly anchored as the foundations. Proving the correctness of an assertion, therefore, amounted to reducing it to the "obvious" validity of the foundation stones.

Euclid's five foundation stones, or "obvious" axioms, were the following:

I. A straight line may be drawn from any point to any other point.

II. A finite straight line may be extended continuously in a straight line.

III. A circle may be described with any center and any radius.

IV. All right angles are equal to one another.

The fifth postulate is unpleasantly complicated, but the general idea is also conveyed by the following formulation (not used by Euclid, whose axioms do not contain the concept *parallel*):

V. Given a line and a point not on that line, there is not more than one line which can be drawn through the point parallel to the original line.

To prove a statement (such as that a construction attains a given object) meant the following to the Greeks: Reduce the statement to one or more of the axioms above, i.e., make the statement as obvious as these axioms are.

It is easily seen that Euclid's axioms are visualizations of the most elementary geometrical constructions, and that they are accomplished by straightedge and compasses. (This is where the straightedge and and compasses come in as an entirely subordinate consideration.) It is evident that if a construction uses more than straightedge and compasses (e.g., a French curve), it can never be reduced to the five

Euclidean axioms, that is, it could never be proved in the eyes of the ancient Greeks.[23] That, and nothing else, was the reason why the Greeks insisted on straightedge and compasses.

Let us illustrate this by a specific example, the squaring of a rectangle. We are required to construct a square whose area is equal to the area of a given rectangle *ABCD*.

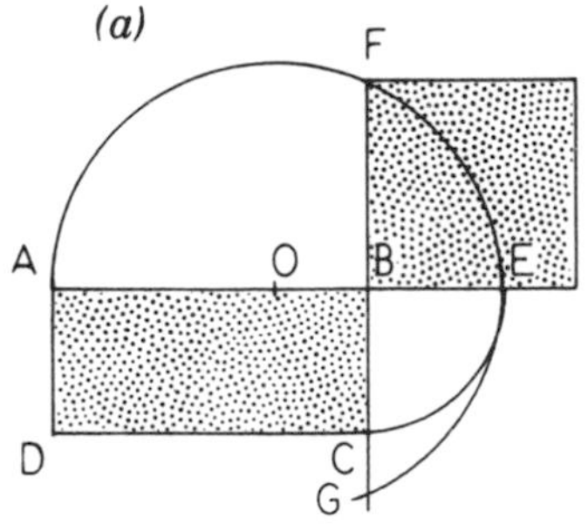

How to square a rectangle.

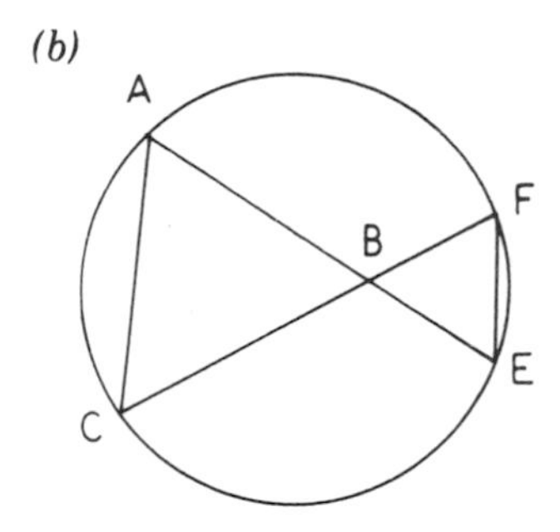

Building stone no. III.35.

Construction (see Figure *a* above): Extend *AB*, and make $BE = BC$. Bisect *AE* to obtain *O*, and with *O* as center, describe a circle with radius *OA*. Extend *CB* to intersect the circle at *F*. Then *BF* is the side of the required square.

Proof: Let *G* be the intersection of *BC* extended and the circle *AFE*. Then by Euclid III.35 (Book III, Theorem 35, to be quoted below),

$$AB \times BE = FB \times BG.$$

But by Euclid III.3,

$$FB = BG$$

and by construction,

$$BE = BC$$

Hence,

$$AB \times BC = FB^2 \quad ,$$

as was to be demonstrated.

The two theorems referred to above are

III.35. If two chords intersect within a circle, the rectangle contained by the segments of one equals the rectangle contained by the segments of the other.

III.3. If the diameter of a circle bisects a chord, it is perpendicular to it; or, if perpendicular to it, it bisects it.

This would usually be all that is required for the proof of the construction, for both theorems are proved by Euclid. However, let us trace the path back to the axioms. Our proof rests, among others, on building stone no. III.35, which says (figure *b* on p. 51) that

$$AB \times BE = CB \times BF.$$

This rests, among others, on the theorem that similar triangles (*CAB* and *BFE*) have proportional sides. To show that these triangles are similar, we have to show, among other things, that the angles *CAB* and *BFE* are equal. This theorem, in turn, rests on the theorem that the angle at the circumference equals half the angle at the center, subtended by the same chord. This, in turn, rests on the theorem that the angles at the base of an isosceles triangle are equal. This again rests on the theorem that if one triangle has two sides and the included angle equal to two sides and the included angle of another, then the two triangles are congruent. And the congruency theorems finally rest almost immediately on the Axioms.

We have traced but a single path from stone no. III.35 down to the foundation stones of Euclid's cathedral. This path has many other branches (branching off where it says "among others" above), and if we had the patience, these could be traced down to the foundation stones, too. So could the paths from stone no. III.3 and from the operations used in the construction (whose feasibility need not be taken for granted).

Thus, the proof that the construction for squaring a rectangle as shown in the figure on p. 51 is correct, is essentially a reduction to Euclid's axioms.

No such reduction is possible with Hippias' construction for squaring the circle. For the reasons given above, it cannot find a resting place in Euclid's cathedral; the paths of proof (such as the derivation given on pp. 41-42) will not lead to Euclid's foundation stones, and Hippias' stone will tumble to the ground.

This does not mean that Hippias' construction is incorrect. Hippias' stone can find rest in a cathedral built on different foundations. What Lindemann proved in 1882 was not that the squaring of the circle (or its rectification, or a geometrical construction of the number π) was impossible; what he proved (in effect) was that it could not be reduced to the five Euclidean axioms.

LET us now take a brief look at the later fate of these five foundation stones, the five Euclidean axioms or postulates. There were a few flaws in them which were later mended, but we have no time for hairsplitting here. We will take a look at a more significant aspect.

The attitude of the ancient Greeks to Euclidean geometry was essentially this: "The truth of these five axioms is obvious; therefore everything that follows from them is valid also."

The attitude of modern mathematics is somewhat different: "If we assume that these axioms are valid, then everything that follows from them is valid also."

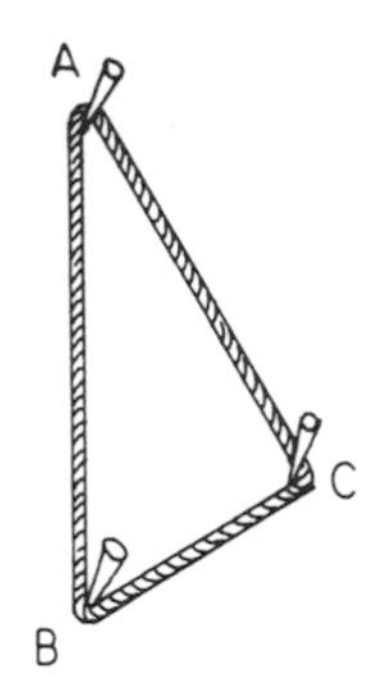

What does "straight" mean?

At first sight the difference between the two seems to be a chicken-hearted technicality. But in reality it goes much deeper. In the 19th century it was discovered that if the fifth postulate (p. 50) was pulled out from under Euclid's cathedral, not all of the building would collapse; a part of the structure (called *absolute geometry*) would remain supported by the other four axioms. It was also found that if the fifth axiom was replaced by its exact opposite, namely, that it *is* possible to draw more than one straight line through a point parallel to a given straight line, then on this strange fifth foundation stone (together with the preceding four) one could build all kinds of weird and wonderful cathedrals. Riemann, Lobachevsky, Bolyai and others built just such crazy cathedrals; they are known as non-Euclidean geometry.

The non-Euclidean axiom may sound ridiculous. But an axiom is unprovable; if we could prove it, it would not be an axiom, for it could be based on a more primitive (unprovable) axiom. We just assume its validity or we don't; all we ask of an axiom is that it does not lead to contradictory consequences. And non-Euclidean geometry is just as free of contradictions as Euclidean is. One is no more "true" than the other. The fact that we cannot draw those parallel lines in the usual way proves nothing.

Nevertheless, some readers may feel that all this is pure mathematical abstraction with no relation to reality. Not quite. Reality is what is confirmed by our experience. It is our experience that when we join two stakes *A* and *B* by a rope, there is a certain configuration of the rope for which the length of rope between *A* and *B* is minimum. We find that configuration of the rope by stretching it, and we call it a straight line. Take a third stake *C* and stretch a rope from *A* to *B* via *C*. Our experience tells us that this rope is longer than the rope stretched between *A* and *B* directly. This experience is in agreement with the Euclidean theorem that the sum of two sides of a triangle

Three who chose to reject Euclid's fifth postulate:
Nikolai Ivanovich LOBACHEVSKI (1792-1856),
Georg Friedrich Bernhard RIEMANN (1826-1866),
and Albert EINSTEIN (1879-1955).

cannot be less than the third side. Euclidean geometry is *convenient* for describing this kind of experience; which is not the same thing as saying it is "universally true."

For there are other experiences for whose description Euclidean geometry is extremely inconvenient. Suppose point A is on this page of the book and point B is on some star in a distant galaxy; then what does "straight" mean? In that case we have no experience with ropes, but we do have experience with light rays. And this experience shows that light rays traveling through gravitational fields do not behave like ropes stretched between stakes. Their behavior is described by Einstein's General Theory of Relativity, which works with non-Euclidean geometry. This is more *convenient* in describing the laws that govern our experience. If we were to express these laws in Euclidean space, they would assume very complicated forms, or alternatively, we would have to revamp all of our electromagnetic theory from scratch (without guarantee of success), and this is not considered worth while (by the few physicists who have given this alternative any thought).

And so the chicken-hearted technicality of saying "if" is neither chicken-hearted nor a technicality.

5

The Roman Pest

Ave Caesar, morituri te salutant!
(Hail Caesar, we who are about to die salute you)
Führer befiehl, wir folgen!
(Führer, command us, we shall follow)

WHILE the quest for knowledge was storming ahead at the University of Alexandria, the ominous clouds of the coming Roman Empire were already gathering. Whilst Alexandria had become the world capital of thinkers, Rome was rapidly becoming the capital of thugs.

Rome was not the first state of organized gangsterdom, nor was it the last; but it was the only one that managed to bamboozle posterity into an almost universal admiration. Few rational men admire the Huns, the Nazis or the Soviets; but for centuries, schoolboys have been expected to read Julius Caesar's militaristic drivel ("We inflicted heavy losses upon the enemy, our own casualties being very light") and Cato's revolting incitements to war. They have been led to believe that the Romans had attained an advanced level in the sciences, the arts, law, architecture, engineering and everything else.

It is my opinion that the alleged Roman achievements are largely a myth; and I feel it is time for this myth to be debunked a little. What the Romans excelled in was bullying, bludgeoning, butchering and blood baths. Like the Soviet Empire, the Roman Empire enslaved peoples whose cultural level was far above their own. They not only ruthlessly vandalized their countries, but they also looted them, stealing their art treasures, abducting their scientists and copying their technical know-how, which the Romans' barren society was rarely able to improve on. No wonder, then, that Rome was filled with great works of art. But the light of culture which Rome is supposed to have emanated was a borrowed light: borrowed from the Greeks and the other peoples that the Roman militarists had enslaved.

There is, of course, Roman Law. They scored some points here, a layman must assume. Yet the ethical substance of our law comes from Jewish Law, the Old Testament; as for the ramifications, the law in English speaking countries is based on the Common Law of the Anglo-Normans. Trial by jury, for example, was an Englishman's safeguard against tyranny, an institution for which he was, and perhaps still is, envied by the people of continental Europe, whose legal codes are based on Roman Law. Even today, this provides for trial by jury only in important criminal cases. Roman Law never had such vigorous safeguards against tyranny as, for example, the Athenian constitution had in the device of ostracism (in a meeting in which not less than 6,000 votes were cast, the man with the highest number of votes was exiled from Athens for 10 years). So I would suspect that what the Romans mainly supplied to our modern lawyers in abundant quantities are the phrases with which they impress their clients and themselves: *Praesumptio innocentiae* sounds so much more distinguished than "innocent until proven guilty," and a maxim like *Ubi non accusator, ibi non judex* shows profound learning and real style. Freely translated, it means "where there is no patrol car, there is no speed limit."

Then there is Roman engineering: the Roman roads, aequeducts, the Colosseum. Warfare, alas, has always been beneficial to engineering. Yet there are unmistakable trends in the engineering of the gangster states. In a healthy society, engineering design gets smarter and smarter; in gangster states, it gets bigger and bigger. In World War II, the democracies produced radar and split the atom; German basic research was far behind in these fields and devoted its efforts to projects like lenses so big they could burn Britain, and bells so big that their sound would be lethal. (The lenses never got off the drawing board, and the bells, by the end of the war, would kill mice in a bath tub.) Roman engineering, too, was void of all subtlety. Roman roads ran absolutely straight; when they came to a mountain, they ran over the top of the mountain as pigheadedly as one of Stalin's frontal assaults. Greek soldiers used to adapt their camps to the terrain; but the Roman army, at the end of a days' march, would invariably set up exactly the same camp, no matter whether in the Alps or in Egypt. If the terrain did not correspond to the one and only model decreed by the military bureaucracy, so much the worse for the terrain; it was dug up until it fitted into the Roman Empire. The Roman aequeducts were bigger than those that had been used centuries earlier in the ancient world; but they were administered with extremely poor knowledge of hydraulics. Long after Heron of Alexandria (1st century A.D.) had designed water clocks, water turbines and two-cylinder water pumps, and had written works on these subjects, the Romans were still describing the performance of their aequeducts in terms of the *quinaria*, a measure of the cross-section of the flow, as if the volume of the flow did not also depend on its velocity. The same unit was used in charging users of large pipes tapping the aequeduct; the Roman engineers failed to realize that doubling the cross-section would more than double the flow of water. Heron could never have blundered like this.

The architecture of the thugs also differs from that of normal societies. It can often be recognized by the megalomaniac style of their public buildings and facilities. The Moscow subway is a faithful copy of the London Underground, except that its stations and corridors are filled with statues of *homo sovieticus*, a fictitious species that stands (or sits on a tractor), chin up, chest out, belly in, heroically gazing into the distance with a look of grim determination. The Romans had similar tastes. Their public latrines were lavishly decorated with mosaics and marbles. When a particularly elaborately decorated structure at Puteoli was dug up by archaeologists in the last century, they thought at first that they had discovered a temple; but it turned out to be a public latrine.[24]

The architecture of the Colosseum and other places of Roman entertainment are difficult to judge without recalling what purpose they served. It was here that gladiators fought to the death; that prisoners of war, convicts and Christians were devoured by as many as 5,000 wild beasts at a time; and that victims were crucified or burned alive for the entertainment of Roman civilization. When the Romans screamed for ever more blood, artificial lakes were dug and naval battles of as many as 19,000 gladiators were staged until the water turned red with blood. The only Roman emperors who did not throw Christians to the lions were the Christian emperors: They threw pagans to the lions with the same gusto and for the same crime — having a different religion.

The apology that the Romans "knew no better at the time" is quite invalid. Like the Nazis, the Romans were not primitive savages, but sophisticated killers, and they certainly knew better from the people they had enslaved. The Greeks vehemently (but unsuccessfully) resisted the introduction of gladiatorism into their country by the Roman overlords; what they must have felt can perhaps be appreciated by the Czechs who, in 1968, watched the Soviet cut-throats amusing themselves by riddling the Czech Museum with machinegun fire. The Ptolemies in Alexandria also went in for pageantry to entertain the people, but they did this without bloodshed and with a sense of humor; one of the magnificent parades staged by Ptolemy II displayed the animals of the king's private zoo, as well as a 180 foot long gilded phallus.[24]

The Romans' contribution to science was mostly limited to butchering antiquity's greatest mathematician, burning the Library of Alexandria, and slowly stifling the sciences that flourished in the colonies of their Empire. The *Naturalis Historia* by Gaius Plinius Secundus[25] (23-73 A.D.) is an encyclopaedic compilation which is generally regarded as the most significant scientific work to have come out of Rome; and it demonstrates the Romans' abysmal ignorance of science when compared to the scientific achievements of their contemporaries at Alexandria, even a century after the Romans had sacked it. For example, Pliny tells us that in India there is a species of men without mouths who subsist by smelling flowers.

The Roman contribution to mathematics was little more than nothing at all.[26] There is, for example, Posidonius (135-51 B.C.), friend and teacher of Cicero and Pompey, who, using a method similar to that of Erastosthenes (p. 46), calculated the circumference of the earth with high accuracy. But if one digs a little deeper, Posidonius' original name is found to be Poseidonios; he was a Syrian who studied in Athens and settled at Rhodes, whence he was sent, in 86 B.C., to Rome as an *envoy*. Poseidonios was therefore as Roman as Euler was Russian. (Euler, as we shall see in Chapter 14, was a Swiss who lived some years in Russia; in Soviet textbooks, he is often referred to as "our great Russian mathematician Eyler.") Poseidonios' value for π must have been accurate to (the equivalent of) several decimal places, for the value he obtained for the circumference of the earth was three centuries later adopted by the great Alexandrian astronomer Ptolemy (no relation to the Alexandrian kings), and this was the value used by Christopher Columbus on his voyage to the New World.

But whatever the value of π used by Poseidonios, it was high above Roman heads. The Roman architect and military engineer Pollio Vitruvius, in *De Architectura* (about 15 B.C.) used the value $\pi = 3\ 1/8$,[27] the same value the Babylonians had used at least 2,000 years earlier.

It is, of course, a simple matter to pick out the bad things in anything that is *a priori* to be run down; to tell the truth but not the whole truth is the basic trick of any propaganda service that has risen above outright lying. Was there, then, nothing good about ancient Rome? Of course there was; it is an ill wind that blows nobody any good. But in a brief background that is getting too far away from π already, I am not concerned with the somebodies to whom Rome may have blown some good; I am only saying that the wind that blew from Rome was an ill wind.

Yet most historians extol the achievements of Rome, and it is only fair to hear some of their reasons. For example:[28]

> Whatever Rome's weaknesses as ruler of empire may have been, it cannot be denied that her conquest of the Western World contributed a great deal to subsequent civilization. It accustomed the Western races to the idea of a world-state, and by *pax romana* (Roman peace) it demonstrated the benefits of a long absence of war, even if the price was the loss of political independence by most of the races of the world.

Simple, is it not? It appears we missed the benefits of *pax germanica* through Winston Churchill and similar warmongers, but all is not lost yet: We still have the chance of *pax sovietica*.

BEFORE Rome became a corrupt empire, it was a corrupt republic.

Across the Mediterranean, in what today is Tunisia, another city, Carthage, had risen and its dominions were expanding along the African coast and into Spain. That provoked Rome's jealousy, and Carthage was defeated in two Punic Wars; but each time she rose again, and in the Third Punic War, after the Carthaginians had withstood a siege against hopeless odds with almost no resources, the Romans captured the city, massacred its population and destroyed the city (146 B.C.)

There had been no rational reason for the Punic Wars. In particular, the Third Punic War was fomented by a group of paranoic hawks in the Roman senate who felt threatened by Carthage's revival. They were led by the pious superpatriot Marcus Porcius Cato (the Elder), who had distinguished himself in the Second Punic War (218-201 B.C.) and who held a number of high offices in the Roman republic, in the course of which he bloodily crushed an insurrection in Spain, raised the rents of the tax-farmers and adjusted the prices of slaves. He vigorously opposed any kind of innovation or reform, strove to stem the tide of Greek refinement, and advocated a return to the strict social life of earlier days; in an age when slaves were branded, flogged and crucified, he was known for the cruelty with which he treated his slaves.[29] No matter what was being discussed in the senate, his speeches would always end with the words *Ceterum censeo Carthaginem esse delendam* (For the rest, I hold that Carthage must be destroyed), a sentence that has been copied in innumerable variations* by people to whom vicious bigots like Cato were presented as examples of the noble Roman spirit.

Such is the background of the Punic Wars, which lead us back to the story of π. During the Second Punic War, the Romans sent an expeditionary force under Claudius Marcellus to Sicily in 214 B.C. For one thing, the king of Syracuse had renewed his alliance with Carthage; for another, the Romans specialized in winning easy victories over small foes.

But this time it was not so easy. Roman brute force, assaulting the city of Syracuse by land and sea, ran into scientific engineering; the engineering that is not bigger, but smarter. The Syracusans had been taught the secret of the lever and of the multiple pulley, and they put it to use in their artillery and marine defenses. The Roman land forces reeled back under the storm of catapult balls, catapult darts, sling bullets and crossbow bolts. The attack by sea fared no better. Syracusan grapnels were lowered from cranes above the cliffs until they caught the bows of Roman ships, which were then hoisted by multiple pulleys until the ships hung vertically and the proud warriors of mighty Rome tumbled into the sea; Roman devices to scale the walls of the city from the ships were battered to pieces by boulders suspended from cranes that swung out over the city walls as the Roman fleet approached. What was then left of the crippled Roman fleet withdrew, and Marcellus hatched a new plan. Under cover of darkness, the Romans sneaked by land to the walls of Syracuse,

* For example, Maria Theresa, empress of Austria (1717-1780), was given the following advice in a note by her personal physician: *Ceterum censeo clitorem Vostris Sanctissimae Majestatis ante coitum excitandam esse.*

thinking that the defenders' catapults could not be used at close quarters. But here they ran into more devilish machines. Plutarch reports that "the wall shot out arrows at all points," and that "countless evils were poured upon them from an unseen source" even after they had fled and tried to regroup. Once more the haughty Roman warriors withdrew to lick their wounds, and Marcellus ranted against this foe "who uses our ships like cups to ladle water from the sea . . . and outdoes all the hundred-handed monsters of fable in hurling so many missiles against us all at once." In the end the invincible Roman legions became so filled with fear that they would run as soon as they saw a piece of rope or wood projecting over the wall. Marcellus had to settle for a siege that was to last the better part of three years.

But, as Bernard Shaw said, God is on the side of the big batallions;[30] and the city finally fell to the Roman cut-throats (212 B.C.), who sacked, plundered and looted it by all the rules of Roman civilization. Inside the city was the 75-year old thinker who had grasped the secret of the lever, the pulley and the principle of mechanical advantage. Plutarch tells us that "it chanced that he was alone, examining a diagram closely; and having fixed both his mind and his eyes on the object of his inquiry, he perceived neither the inroad of the Romans nor the taking of the city. Suddenly a soldier came up to him and bade him follow to Marcellus, but he would not go until he had finished the problem and worked it out to the proof."

"Do not touch my circles!" said the thinker to the thug. Thereupon the thug became enraged, drew his sword and slew the thinker.

The name of the thug is forgotten.

The name of the thinker was Archimedes.

6

Archimedes of Syracuse

Greek scholars are privileged men; few of them know Greek, and most of them know nothing else.

George Bernard SHAW
(1856-1950)

WHEN Newton said "If I have seen further than others, it is because I stood on the shoulders of giants," one of the giants he must have had in mind was Archimedes of Syracuse, the most brilliant mathematician, physicist and engineer of antiquity.

Little is known about his life. He was born about 287 B.C. in Syracuse, the son of the astronomer Pheidias, and apparently spent most of his life in Syracuse. He studied at the University of Alexandria either under Euclid's immediate successors, or perhaps under Euclid himself. He was a kinsman and friend of Hieron II, king of Syracuse, for whom he designed the machines of war used against the Roman aggressors, and whose crown was involved in the discovery of the law of upthrust that bears his name. Hieron suspected (correctly) that his crown was not pure gold and asked Archimedes to investigate without damaging the crown. Archimedes is said to have pondered the problem while taking a bath, and to have found the answer as he observed the water level rising on submerging his body into the bath. Shouting ***Heureka!*** (I have found it), says this legend, he ran naked through the streets of Syracuse to tell Hieron of his discovery.

His book *On Floating Bodies* goes far beyond Archimedes' Law, and includes complicated problems of buyoancy and stability. Likewise, *On the Equilibrium of Planes* goes beyond the principle of the lever and solves complicated problems such as finding the center of gravity of a parabolic segment. In these, as in his other works, Archimedes used the Euclidean approach: From a set of simple postulates, he deduced his propositions with unimpeachable logic. As the first writer who consistently allied mathematics and physics, Archimedes became the father of physics as a science. (Aristotle's *Physics* was published a century earlier, but it is only a long string of unfounded speculations, totally void of any quantitative relations.)

ARCHIMEDES OF SYRACUSE
(ca. 287–212 B.C.)

Archimedes was also the first scientific engineer, the man searching for general principles and applying them to specific engineering problems. His application of the lever principle to the war machines defending Syracuse are well known; yet he also applied the same principle to find the volume of the segment of a sphere by an unusually beautiful balancing method about which we shall have more to say later in this chapter. He used the same method to determine the volumes of other solids of revolution (ellipsoid, paraboloid, hyperboloid) and to find the center of gravity of a semicircle and a hemisphere. It is not known how many of Archimedes' works have been lost (one of the most important, *The Method*, came to light only in 1906), but his extant books, including *On Spirals*, *On the Measurement of the Circle*,

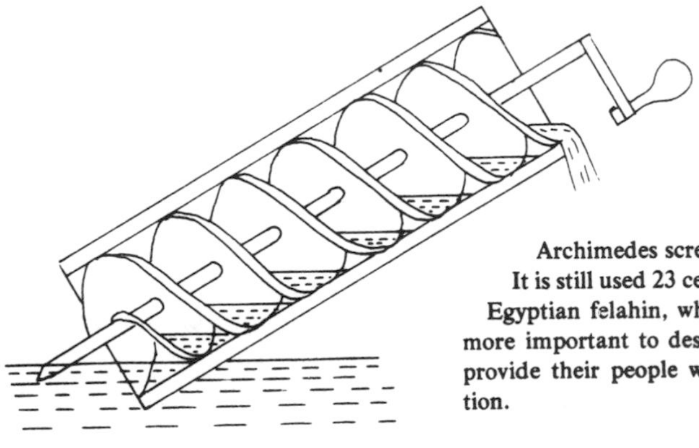

Archimedes screw or helical pump. It is still used 23 centuries later by the Egyptian felahin, whose rulers think it more important to destroy Israel than to provide their people with modern irrigation.

Quadrature of the Parabola, On Conoids and Spheroids, On the Sphere and Cylinder, Book of Lemmas and others, are unmatched by anything else produced in antiquity.

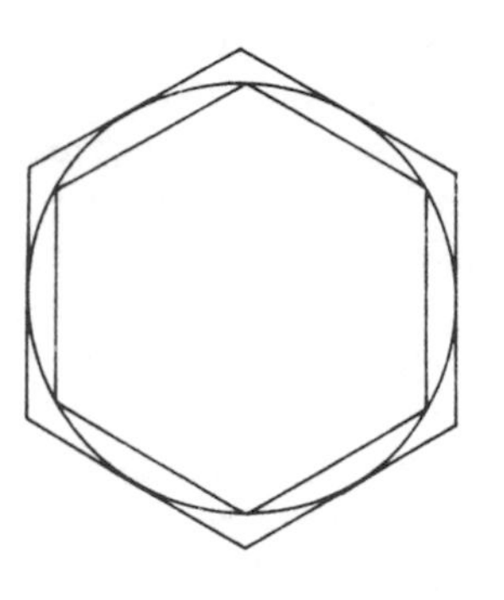

Archimedes' method of calculating π.

Not only because of the marvelous results contained in these books. But also because Archimedes was a pioneer of method. He took the step from the concept of "equal to" to the concept of "arbitrarily close to" or "as closely as desired" (which Euclid had enunciated, but not actively used) and thus reached the threshold of the differential calculus, just as his method of squaring the parabola reached the threshold of the integral calculus (some consider he crossed it).

He was also the first to give a method of calculating π to any desired degree of accuracy. It is based on the fact that the perimeter of a regular polygon of n sides inscribed in a circle is smaller than the circumference of the circle, whereas the perimeter of a similar polygon circumscribed about the circle is greater than its circumference (see figure above). By making n sufficiently large, the two perimeters will approach the circumference arbitrarily closely, one from above, the other from below. Archimedes started with a hexagon, and progressively doubling the number of sides, he arrived at a polygon of 96 sides, which yielded

$$3\tfrac{10}{71} < \pi < 3\tfrac{1}{7} \tag{1}$$

or in decimal notation,

$$3.14084 < \pi < 3.142858 \tag{2}$$

That Archimedes did this without trigonometry, and without decimal (or other positional) notation is an illustration of his tenacity (see Heath's translation of *On the Measurement of the Circle*, Proposition 3 and following). However, we shall use both of these to go through the calculation.

If $\theta = \pi/n$ is half the angle subtended by one side of a regular polygon at the center of the circle, then the length of the inscribed side is

$$i = 2r\sin\theta \tag{3}$$

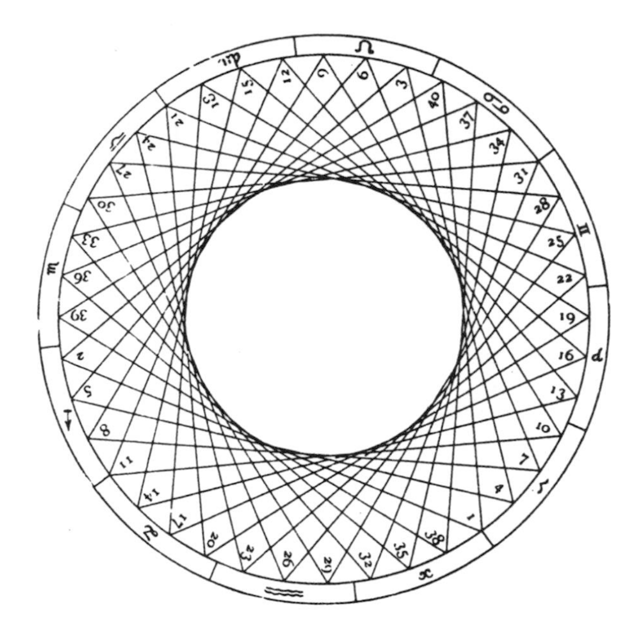

A regular polygon of 40 sides. No internal circle has been drawn.[31]

and that of a circumscribed side is

$$c = 2r\tan\theta \tag{4}$$

For the circumference C of the circle we therefore have

$$ni < C < nc$$

or dividing by $2r$,

$$n\sin\theta < \pi < n\tan\theta . \tag{5}$$

If the original number of sides n is doubled k times, this yields

$$2^k n\sin(\theta/2^k) < \pi < 2^k n\tan(\theta/2^k) \tag{6}$$

and by making k sufficiently large, the lower and upper bounds will approach π arbitrarily closely.

Archimedes did not, of course, use trigonometric functions; however, for $n = 6$, he had $\sin\theta = 1/2$, $\tan\theta = \sqrt{(1/3)}$ by Pythagoras'

Theorem, and the remaining functions in (6) can be obtained by the successive use of the half-angle formulas (which correspond to finding proportions in right-angled triangles). For $k = 4$, the two polygons will have 96 sides, and this will lead to the limits (1), if the square roots involved in the half-angle formulas are approximated by slightly smaller rational numbers for the lower limit, and by slightly larger rational numbers for the upper limit.

This again is easier described in modern terminology than actually done without trigonometry or a decimal system for the accompanying arithmetic. The half-angle formulas most closely resembling Archimedes' procedure are

$$\cot(\theta/2) = \cot\theta + \operatorname{cosec}\theta$$
$$\operatorname{cosec}^2(\theta/2) = 1 + \cot^2(\theta/2),$$

which enabled him (in effect) to find $\cot(\theta/2)$ and $\operatorname{cosec}(\theta/2)$ from $\cot\theta$ and $\operatorname{cosec}\theta$. For a hexagon, Archimedes approximated $\sqrt{3}$ by the slightly smaller value 265/153; a 12-sided polygon already involved him in the ratio $\sqrt{(349450)} : 153$, which he simplified to $591\ {}^1/_8 : 153$; and the final 96-sided polygon involved a square root of a number that in the decimal system had ten digits! How he managed to extract his square roots with such accuracy, always taking care to keep slightly on the small or large side as demanded by the bounds, is one of the puzzles that this extraordinary man has bequeathed to us.

But it appears that Archimedes went even further. Heron of Alexandria, in his *Metrica* (about 60 A.D., but not disovered until 1896), refers to an Archimedes work that has since been lost, where Archimedes gives the bounds

$$211875 : 67441 < \pi < u$$

or

$$3.14163 < \pi < u,$$

where the upper limit u, in the copy of Heron's work found in Constantinople in 1896, is given as

$$u = 197888 : 62351 \quad (= 3.1738);$$

however, this is evidently an error that must have crept in during the transcription of the copy, for this is far coarser than the upper bound 3 1/7 found by Archimedes earlier. Heron adds "Since these numbers are inconvenient for measurements, they are reduced to the ratio of the smaller numbers, namely 22 : 7."

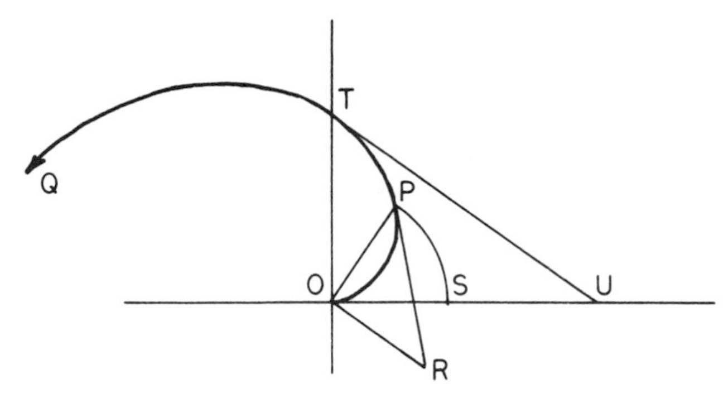

Rectification of the circle by the Archimedean Spiral

The bounds attained by Archimedes in the *Measurement of the Circle* were

$$3\ 10/71 < \pi < 3\ 1/7$$

or in decimal notation,

$$3.140845\ldots < \pi < 3.142857\ldots \qquad (7)$$

Archimedes also showed that a curve discovered by Conon of Alexandria could, like Hippias' quadratrix, be used to rectify (and hence square) the circle. The curve is today called the Archimedean Spiral; it is defined as the plane locus of a point moving uniformly along a ray while the ray rotates uniformly about its end point. It will thus be traced by a fly crawling radially outward on a turning phonograph record.

Let P be any point on the spiral (see figure above), and let the tangent at P intersect the line OP at R. Archimedes showed in his book *On Spirals* that the segment OR (i.e. the polar subtangent at the point P) equals the length of the circular arc PS described with radius OP, where S is the intersection with the initial ray OU. It follows that OU equals ¼ of the circumference of a circle with radius OT, so that

$$2OU : OT = \pi \qquad (8)$$

and the area of the triangle OTU is

$$\tfrac{1}{2} OT \times QU = \tfrac{1}{2} OT \times \tfrac{1}{2}\pi OT = \tfrac{1}{4}\pi \times OT^2$$

whence

area of circle with radius OT = 4 times area of ΔOTU, and the circle is squared once more, though not to the liking of Greek geometry; the objections are the same as in the case of Hippias' quadratrix.

Archimedes used a double *reductio ad absurdum* to prove that the segment *OR* and the arc *PS* were equal in length. We shall prove it more quickly by differential geometry. Let $OP = \rho$ and the angle *SOP* $= \theta$. Then from the definition of the Archimedean Spiral,

$$\rho = k\theta \tag{9}$$

where *k* is a constant equal to the ratio of angular and linear velocities. As shown in textbooks of differential geometry, the length of a subtangent in polar coordinates is

$$OR = -\rho^2(\theta) / \rho'(\theta) \tag{10}$$

and since $\rho'(\theta) = k$, we have from (9), (10), and the figure

$$OR = k\theta^2 = \rho\theta = \text{arc}\, PS, \tag{11}$$

as was to be demonstrated. In particular, for $\theta = \pi/2$,

$$OU = \tfrac{1}{2}\pi \times OT \tag{12}$$

SUCH were some of the contributions of antiquity's greatest genius to the history of π and squaring the circle. Though later investigators found closer numerical approximations, Archimedes' polygonal method remained unsurpassed until 19 centuries later an infinite product and an infinite continued fraction were found in England just before the discovery of the differential calculus led to a totally new approach to the problem.

In physics, no one came near Archimedes' stature for 18 more centuries, until Galileo Galilei dared to challenge Aristotle's humbug.

Shadows of that humbug are with us even now. There is hardly a history of mathematics that does not apologize for Archimedes for getting his hands dirty with experimental work. As late as 1968, we are told that "he placed little value in his mechanical contrivances," and (1965) that "he regarded his contrivances and inventions as sordidly commercial." Similar comments will be found in almost all

Proposition 2.

If a right segment of a paraboloid of revolution whose axis is not greater than $\frac{3}{4}p$ (where p is the principal parameter of the generating parabola), and whose specific gravity is less than that of a fluid, be placed in the fluid with its axis inclined to the vertical at any angle, but so that the base of the segment does not touch the surface of the fluid, the segment of the paraboloid will not remain in that position but will return to the position in which its axis is vertical.

Let the axis of the segment of the paraboloid be AN, and through AN draw a plane perpendicular to the surface of the fluid. Let the plane intersect the paraboloid in the parabola BAB', the base of the segment of the paraboloid in BB', and the plane of the surface of the fluid in the chord QQ' of the parabola.

Then, since the axis AN is placed in a position not perpendicular to QQ', BB' will not be parallel to QQ'.

Draw the tangent PT to the parabola which is parallel to QQ', and let P be the point of contact*.

[From P draw PV parallel to AN meeting QQ' in V. Then PV will be a diameter of the parabola, and also the axis of the portion of the paraboloid immersed in the fluid.

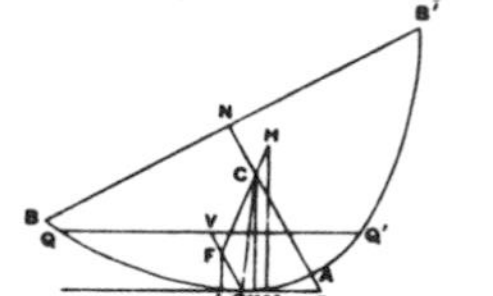

Let C be the centre of gravity of the paraboloid BAB', and F that of the portion immersed in the fluid. Join FC and produce it to H so that H is the centre of gravity of the remaining portion of the paraboloid above the surface.

Then, since $AN = \frac{3}{2}AC$*,

and $AN \ngtr \frac{3}{4}p$,

it follows that $AC \ngtr \frac{p}{2}$.

ARCHIMEDEAN SCIENCE[32]

8

We shall now proceed to set forth the fact that it is possible for a motion, which is one and continuous, to be infinite, and the fact that this motion is circular.

Now everything in locomotion is moved with a motion which is either circular or rectilinear or a blend of the two; so if one of the first two kinds of motion is not continuous,[1] neither can the composite of the two be.[2] It is clear that a thing in locomotion along a finite straight line cannot proceed continuously, for it must turn back;[3] and that which turns back along a straight line has contrary motions, for an upward locomotion is contrary to a downward locomotion, a forward locomotion is contrary to a backward one, and that to the left is contrary to that to the right, and this is so since the corresponding places are pairs of contraries. Previously we have defined a motion as being one and continuous if it is of one thing and during one time and without difference in kind.[4] For there are three things here[5]—(a) that which is in motion, e.g., a man or a divine being;[6] (b) the whenness of the motion, i.e., the [individual] time; and (c) that in which [i.e., the category], and this is either a place or an affection or a form[7] or a magnitude. Now contraries differ in species[8] and are not one; and the differentiae of place are those which have just been stated. A sign that the motion from A to B is contrary to that from B to A is the fact that, if occurring simultaneously, they stop or cancel each other.[9] And it is likewise with motions along a circle, e.g., that from A going toward B is contrary to that from A going the other way toward C, for these too stop even if they are continuous, and there is no return, and this is because contraries destroy or obstruct each other;[10] but a lateral motion is not contrary to an upward motion.[11] That a rectilinear motion cannot be continuous is most evident from the fact that the object must stop before turning back. This is so not only in the case of a rectilinear motion, but also if the object were to go around a circle. (For to have a circular locomotion and to go around a circle are not the same: In the first case, the motion is connected; in the second, the object must come to the starting point but then turn back.)[12]

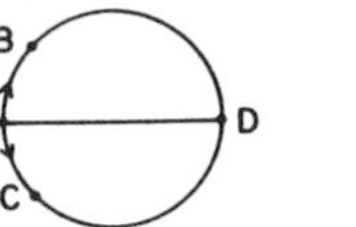

ARISTOTELIAN PRATTLE[33]

books on the subject, bearing out Philip Guedella's remark "History repeats itself; historians repeat each other." The repetitious myth was started by Plutarch, who in the 1st century A.D. wrote that

> Regarding the business of mechanics and every utilitarian art as ignoble and vulgar, he gave his zealous devotion only to those subjects whose elegance and subtlety are untrammeled by the necessities of life.

Now Plutarch could not possibly have known what Archimedes regarded as ignoble and vulgar; his guess was as good as yours or mine. But just as for 19 centuries historians have been echoing Plutarch, so Plutarch echoed the attitude of Plato and Aristotle, the fathers of intellectual snobbery. They taught that experimentation was fit only for slaves, and that the laws of nature could be deduced merely by man's lofty intellect; and Aristotle used his lofty intellect to deduce that heavier bodies fall to the ground more rapidly; that men have more teeth than women; that the earth is the center of the universe; that heavenly bodies never change; and much more of such wisdom, for he was a very prolific writer. As a matter of fact, Aristotle was defeated on his own grounds, by sheer intellectual deduction unaided by experimental observation. Long before Galileo Galilei dropped the wooden and leaden balls from the leaning tower of Pisa, he asked the following question: "If a 10 lb stone falls ten times as fast as a 1 lb stone, what happens if I tie the two stones together? Will the combination fall faster than the 10 lb stone because it weighs 11 lbs, or will it fall more slowly because the 1 lb stone will retard the 10 lb stone?"

Plutarch's comment on Archimedes' attitude to engineering is a concoction with no foundation other than Aristotelian snobbery. Archimedes' book *The Method* was discovered only in the present century, and it sheds some interesting light on the point. A *reductio ad absurdum* is to this day a popular method of proof; it does, however, have a drawback, and that is that one must know the result beforehand in order to know the alternative that is to be disproved. How this result was originally derived does not transpire from the proof. Since Archimedes very often used *reductiones ad absurdum*, historians have often marveled by what method he originally discovered his wonderful results.

It turns out that in many cases he did this by a method of analogy or modeling (as it is called today), based on the principles of none other than the ignoble, vulgar and commercially sordid contrivances that he is alleged to have invented with such reluctance. In *The*

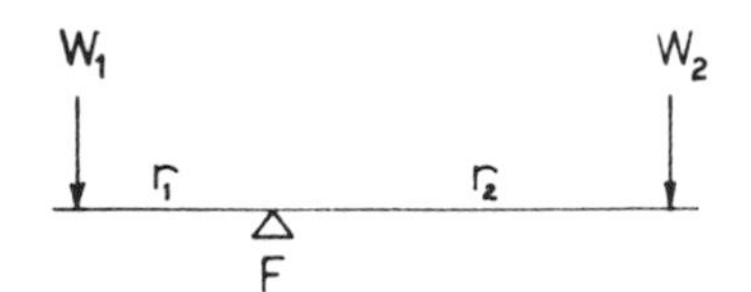

The law of the lever. $W_1r_1 = W_2r_2$, where W_1, W_2 are weights (or other forces).

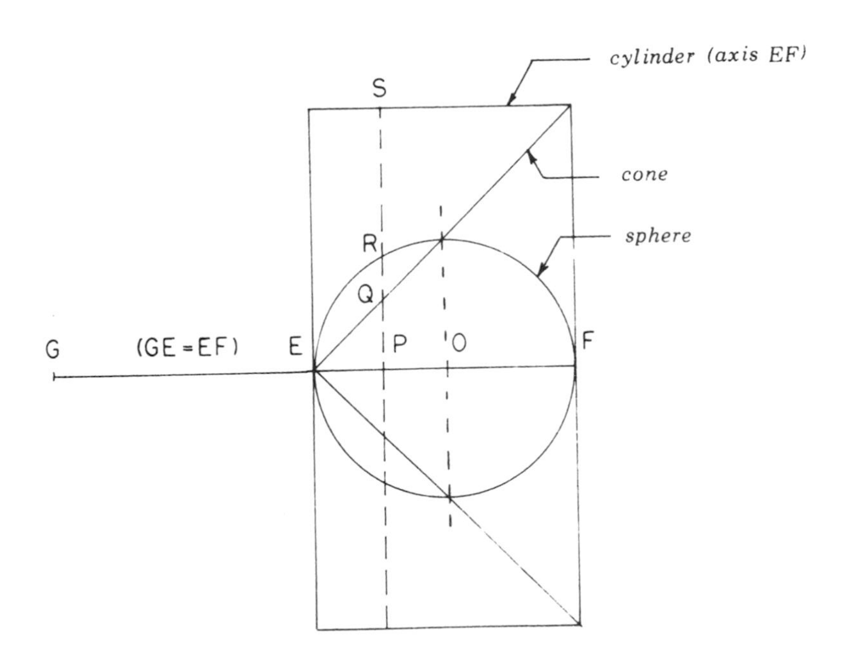

The lever principle applied to geometry. A plane *PS*, perpendicular to *GF* at any point *P* will cut the sphere, cone and cylinder in circles with radii *PR*, *PQ* and *PS*, respectively. Archimedes proved that the first two circles (their weights proportional to their areas) placed on the lever *GEF* with fulcrum *E* would exactly balance the third circle at *P*. From this he derived the volume of a spherical segment, as well as the volume of a whole sphere ($4\pi r^3/3$).

Method, he writes to Erastosthenes, the Librarian at Alexandria whom we have already met (p. 46):[34]

> Archimedes to Erastosthenes, greetings.
> I sent you on a former occasion some of the theorems discovered by me, merely writing out the enunciations and inviting you to discover the proofs, which at the time I did not give. [. . .] The proofs then of these theorems I have written in this book and now send to you. Seeing moreover in you, as I say, an earnest student, [. . .] I thought fit to write out for you and explain in detail in the same book the peculiarity of a certain method, by which it will be possible for you to investigate some of the problems in mathematics by means of mechanics. This procedure is, I am persuaded, no less useful even for the proofs of the theorems themselves; for certain things first became clear to me by a mechanical method, although they had to be demonstrated by geometry afterwards because their investigation by the said method did not furnish an actual demonstration. But it is of course easier, when we have acquired, by the method, some knowledge of the questions, to supply the proof than it is to find it without any previous knowledge.

A vivid example of this method is Archimedes' application of the lever principle to derive the volume of a spherical segment as well as that of a whole sphere, as indicated in the figure on p. 71. So highly did Archimedes value this discovery that he asked for a sphere inscribed in a cylinder to be engraved in his tombstone, and this was done, for though the tombstone has been lost, we have a description of the tomb by Cicero, who visited it in the 1st century A.D. during his office as quaestor in Sicily.

It is not without interest to recall how *The Method* was recovered. It was found in 1906 in Constantinople on a palimpsest, that is, a parchment whose original text has been washed off (to salvage the parchment) and replaced by different text. If the original has been washed off imperfectly, it can be restored by special photography. In this case, the original text was a 10th century copy of some known Archimedean works, but also including the only surviving text of *The Method*. The mediaeval zealots did not always, like the Bishop of Yucatan or the Crusaders at Constantinople, burn scientific books as work of the devil. Sometimes they would only wash off the text for the sake of the parchment, so that they might besmirch it with their superstitious garbage.

7

DUSK

> *Theodotus:* What is burning there is the memory of mankind.
> *Caesar:* A shameful memory. Let it burn.
>
> George Bernard Shaw,
> *Caesar and Cleopatra*

THE University of Alexandria remained the intellectual center of the ancient world even after its library had been burned during the occupation by the Roman hordes. The details of that catastrophe are roughly as follows. When Caesar occupied Alexandria in 48 B.C., the Quisling Queen Cleopatra (a far cry from Hollywood's puerile fantasies) offered him not only her bed, but also the Library. What made a thug like Julius Caesar accept the offer (of the library) is not altogether clear, but he helped himself to a large part of the rolls, which were readied for shipment to Rome. Then the Alexandrians, in an early instance of campus unrest, rose against Caesar and Cleopatra. They were, of course, crushed by Roman might, and the shipment of books for Rome, or the remaining library, or both, were burned in the fighting.

When Cleopatra bedded with the next Roman war lord Marcus Antonius (Mark Anthony), he compensated her with Roman generosity: He stole the 200,000 roll library of Pergamon and graciously presented it to Cleopatra.

The remaining story of the Library of Alexandria is typical for what was in store for science at the hands of political rulers and religious fanatics in the years to come. It was again damaged when the Roman emperor Aurelianus quelled an Egyptian revolt in 272 A.D., and again in 295 when the emperor Diocletian suppressed another revolt. In 391, a Christian mob led by the fanatical Bishop Theophilus destroyed the Temple of Serapis, where some of the books were kept. Another zealous bishop, Cyril, led a Christian mob against the astronomer and mathematician Hypatia; as a devotee of pagan learning, she was hacked to death in 415, and her death marks the end of Alexandria as a center of mathematical learning. The end came in 646, when the Arabs captured the city. It is related that their general Amr ibn-al-As wrote to his Khalif to ask what to do with the books of the library. The reply was that if the books agreed with the Koran, they were superfluous; if they disagreed, they were pernicious. So they were burned.

Many famous mathematicians worked in the Greek world under Roman rule. Ptolemy, the great astronomer, worked in Alexandria in 139-161 A.D. He used the value

$$\pi = 3^{17}/_{120} = 3.14167,$$

which he may have taken over from Apollonius of Perga, an outstanding mathematician some 30 years younger than Archimedes. Eutokios, in the 5th century A.D., comments that Apollonius refined Archimedes' method, but the book he refers to has been lost.

There were others. Heron (1st century A.D.), Diophantus (3rd century), Pappus (late 3rd century), all of Alexandria, were famous mathematicians, but none of them appears to have contributed significantly to the history of π. This was the "Silver Age" of Greek mathematics; still high above anything Rome ever squinted at, but well below the golden age of Euclid, Archimedes and Apollonius.

But already Greek mathematics, like the other sciences, were slowly dying under the cold breath of Rome; it was getting dusk, and the stage was set for a disaster even greater than that of the Roman Empire.

In this age of dusk, when Rome began to change from ferocity to decadence, and when science was beginning to drown in the oncoming flood of mysticism, superstition and dogma, we find the curious figure of Nehemiah, a Rabbi and mathematician, who is perhaps not very significant in the history of mathematics, but who is symbolic for this age, for he made a gallant attempt to reconcile science with religion.

It has been remarked before (p. 16) that the biblical verse I Kings vii, 23 was a curious pebble on the road to the confrontation between science and religion. Nehemiah noticed this pebble, picked it up and contemplated it.

Nehemiah was a Hebrew Rabbi, scholar and teacher who lived in Palestine and wrote about 150 A.D., after the last Judean revolt against the Romans (132-135) led by Bar-Kokba and resulting in the Diaspora of the Jews. Nehemiah was the author of the Mishnat ha-Middot, the earliest Hebrew geometry known to us. Not all of it has been preserved; but it deals with the elements of plane and solid geometry, and with the measurements and construction of the Tabernacle. Our concern here is only with matters touching on π, in particular, the Old Testament verse (I Kings vii, 23, also 2 Chronicles iv,2),

> Also, he made a molten sea ten cubits from brim to brim, round in compass, and five cubits the height thereof; and a line of thirty cubits did compass it round about.

which implies $\pi = 3$.

Nehemiah writes in his textbook:

> The circle has three aspects: the circumference, the thread and the roof. Which is the circumference? That is the rope surrounding the circle; for it is written: *And a rope of thirty cubits did encompass it round about.* And the thread? That is the straight line from brim to brim; for it is written: *From brim to brim.* And the roof itself is the area.
>
> [. . .]
>
> And if you want to know the circumference all around, multiply the thread into three and one seventh . . .

That is the Archimedean value, $\pi = 3\ 1/7$. But Nehemiah was also a Rabbi; so how does he get round I Kings vii, 23? Like this:

> Now it is written: *And he made the molten sea of ten cubits from brim to brim, round in compass,* and yet its circumference is thirty cubits, for it is written: *And a line of thirty cubits did compass it round about.* What is the meaning of the verse *And a line of thirty cubits* and so forth? Nehemiah says: Since the people of the world say that the circumference of a circle contains three times and one seventh of the thread, take off that one seventh for the thickness of the walls of the sea on the two brims, then there remain *thirty cubits did compass it round about.*

Hats off to the crafty old fox! The "people of the world" say $\pi = 3\ 1/7$, but the scriptures say $\pi = 3$; so you measure the *inner* circumference of the walls, whereas the diameter is measured from *outer* rim to *outer* rim; and the thickness of the walls, you blockheads, makes up exactly for that secular one seventh! Certainly the Rabbi had more wits than the dogmatic commentators of the Bible in Germany in the 18th century; crudely ignoring the description "round in compass," they claimed that the molten sea must have been hexagonal.[35]

Nevertheless, the dear Rabbi swindled quite brazenly, for the width of the molten sea walls is given three verses further on (I Kings vii, 26):

> And it was an hand breadth thick, and the brim thereof was wrought like the brim of a cup, with flowers of lilies; it contained two thousand baths.[36]

This was the age of dusk, when it was still possible to attempt compromising between science and religion. No such compromise was tolerated in the mediaeval night. Whoever tampered with what the Bible said, risked torture chamber and the stake.

BEFORE we go on into that night, let us pause to admire the ancients for their tenacity in tackling mathematical problems without the benefit of algebraic symbolism (which was introduced much later by the Arabs). Nehemiah, for example, states the area of a circle as follows:

> If one wants to measure the area of a circle, let him multiply the thread (diameter) into itself and throw away from it the one seventh and the half of one seventh; the rest is the area, its roof.

That is, the area is

$$A = d^2 - d^2/7 - d^2/14,$$

which equals $(3\ 1/7) \times (d/2)^2$, so that if the Archimedean value $\pi = 3\ 1/7$ is accepted, the formula is correct.

There was also no single symbol, such as π, for the circle ratio. In mediaeval Latin, it was described by the following mouthful: *quantitas, in quam cum multiplicetur diameter, proveniet circumferentia* (the quantity which, when the diameter is multiplied by it, yields the circumference), and this clause was inserted in the long sentences that stated the equivalent of our formulas; for example, the area of a circle was given as follows:

Multiplicatio medietatis diametri in se ejus, quod proveniet, in quantitatem, in quam cum multiplicatus diameter provenit circumferentia, aequalis superficies circuli. (Multiplication of half the diameter with itself and of that which results by the quantity, which when the diameter is multiplied by it, yields the circumference, equals the area of the circle.)

This monster sentence states (correctly) that

$$[(d/2) \times (d/2)] \times \pi = A.$$

Perhaps the Greeks made such enormous progress in mathematics because their geometry steered clear of numerical calculations, and thus did not get bogged down in expressing algebraic relations intelligibly. Euclid's statements that

In a circle, equal arcs subtend equal chords

or

If corresponding angles of two triangles are equal, then corresponding sides are proportional

have not been improved on in 2,200 years.

8

NIGHT

What ye have done unto the last of my brethren, ye have done unto me
JESUS OF NAZARETH

THE fall of Rome in 476 A.D. marks the time when the literate barbarians of Rome were replaced by the illiterate barbarians of Germany, and this event is generally considered as the beginning of the Middle Ages.

Disaster was followed by catastrophe: The Roman Empire was followed by the Roman Church. In the Eastern or Byzantine Empire, the Roman Empire continued hand in hand with the Eastern Orthodox church for a thousand more years.

What made the bloody tyrants of the time accept Christianity is debatable. Perhaps they welcomed a religion that would teach their subjects to turn the other cheek, and that promised salvation for humility and eternal damnation for rebellion. Perhaps they were impressed by the unparalleled organization and iron discipline of the early Christians. What is certain is that once the mediaeval kings and emperors had adopted Christianity as the official faith, their executioners saw to it that nothing else was allowed.

And "nothing else," of course, included science, the staunchest opponent of the Supernatural.

In the early days of civilization, science and religion were in the same hands, in the hands of the priests, who had a monopoly on learning. In Egypt, Babylon, the Yucatan, and everywhere else in the Belt, they were the timekeepers and surveyors, the astronomers and geometers. They soon learned that knowledge was power: power over the gullible and over the uneducated. Five thousand years ago, the Chaldaean priests knew how to predict an eclipse, and they would impress the uneducated with much mumbo-jumbo about the impending catastrophe. The Egyptian priests had nilometers in their temples, instruments that communicated with the Nile by secret, underground channels; they learned to predict the rising and falling of the Nile, and they would impress the uneducated with much much mumbo-jumbo about the imminent floods or recessions of the Sacred River.

Mathematics and mumbo-jumbo remained allied for a long time. Pythagoras, for example, was a mathematician to a much lesser degree than he was a mumbo-jumboist. He taught that one stood for reason, two for opinion, three for potency, four for justice, five for marriage, seven held the secret of health, eight the secret of marriage, the even numbers were female, the odd numbers male, and so on. "Bless us, divine number," the Pythagoreans would incant to the number four, "who generatest Gods and men, O holy *tetraktys*, that containest the root and source of the eternally flowing creation."[37] Immerged in this mumbo-jumbo, Pythagoras discovered that $3^2 + 4^2 = 5^2$, and that there were other triplets of numbers x, y, z that satisfy the relation $x^2 + y^2 = z^2$. Translated into geometry, this led him to the theorem which is quite unjustly called after him, for it had been known to every rope-stretcher (surveyor) from the Nile to the Yang-Tse Kiang for a thousand years before Pythagoras' witchcraft.[38]

But witchcraft is easier than mathematics; specialization set in, and mumbo-jumbo became more and more separated from mathematics. Those interested in knowledge took the path to science; others took the road to the occult, the mystic and the supernatural. Thousands of cults and religions developed. The vast majorities of these did not recruit converts by force; not even the Roman enslavers forced their religion on the peoples they had conquered. But in three major cases religion became militant, and its fanatics disseminated it by fire and sword, plunging large parts of humanity into prolonged nightmares of horror. They were mediaeval Christianity, mediaeval Islam, and modern communism. For communism is a religion, too, with its church (party), its priests, its scriptures, its rituals, its dogmas, and its liturgy; the communist claims about dialectical materialism being a science are just so much more mumbo-jumbo. Communism uses science for its own ends; Muslim science, especially mathematics, flourished (on the whole) under the rule of the Mohammedan fanatics; but mediaeval Christianity, in many ways the most terrible of the three catastrophies, persecuted science with torture chamber and the stake, and almost succeeded in extinguishing it completely.

Scientific research, wrote Tertullian (160-230), had become superfluous since the Gospel of Jesus Christ had been received. Tomaso d'Aquino (St. Thomas, 1225-1274) wrote that there was no conflict between science and religion; but by "science" he understood Aristotle's tiresome speculations. The mediaeval Church never made St. Thomas' mistake of confusing the two: It promoted Aristotle to some kind of pre-Christian saint, but (except for a very modest amount of third-rate work within its own cloisters) persecuted science almost wherever it appeared.

Scientific works and entire libraries were set to the torch kindled by the insane religious fanatics. We have already mentioned the Bishop of Yucatan, who burned the entire native literature of the Maya in the 1560's, and Bishop Theophilus, who destroyed much of the remnants of the Library of Alexandria (391). The Christian Roman emperor Valens ordered the burning of non-Christian books in 373. In 1109, the crusaders captured Tripoli, and after the usual orgy of butchery typifying the crusades (though this one did not yet include the murderous Teutonic Knights), they burned over 100,000 books of Muslim learning. In 1204, the fourth crusade captured Constantinople and sacked it with horrors unparalleled even in the bloody age of the crusades; the classical works that had survived until then were put to the torch by the crusaders in what is generally considered the biggest single loss to classical literature. In the early 15th century, Cardinal Ximenes (Jimenez), who succeeded Torquemada as Grand Inquisitor and was directly responsible for the cruel deaths of 2,500 persons, had a haul of 24,000 books burned at Granada.*

In 1486, Torquemada sentenced the Spanish mathematician Valmes to be burned at the stake because Valmes had claimed to have found the solution of the quartic equation. It was the will of God, maintained the Grand Inquisitor of the Holy Office of the Inquisition Against Heretical Depravity, that such a solution was inaccessible to human understanding.[39]

Not only scientific theory was condemned as the work of the devil. The devil also seems to have known a lot more about navigation than the bloodthirsty Men of God. Many (perhaps most) ships sailing the

* In a widely used biographical dictionary, we read that "His munificence as a patron of religion, of letters and of art deserves the highest praise."

Mediterranean in the Middle Ages had Jewish navigators, for the Christian captains and crews were not supposed to meddle with the devilish science of mathematics.[40] In the 10th century, Raud the Strong, a Viking chieftain, escaped the fanatical Christianizing king of Norway Olaf Trygvasson by sailing into the wind (i.e., maintaining a zig-zag course whose average advances against the wind); the pious king, who was better acquainted with witchcraft than with the triangle of forces, thereupon accused Raud of being in alliance with the devil, and when he finally caught him, he had him killed by stuffing a viper down his throat.[41]

GIORDANO BRUNO
(1548-1600)

The Middle Ages are usually considered to have ended with the fall of Constantinople (1453) or the discovery of the New World (1492); but the insane persecution of science continued well beyond that time, and it is difficult to give a date when it ceased. For more than a century longer, the Church tolerated no deviation from the literal word of the Bible, or from the teachings of its idol Aristotle. In 1600, Giordano Bruno was burned alive in Rome for claiming that the earth moves round the sun. In 1633, the 70-year old Galileo Galilei went through the torture chambers of the Inquisition until he was willing to sign that

> I, Galileo Galilei, . . . aged seventy years, being brought personally to judgement, and kneeling before you Most Eminent and Most Reverend Lords Cardinals, General Inquisitors of the universal Christian republic against depravity . . . swear that . . . I will in future believe every article which the Holy Catholic and Apostolic Church of Rome holds, teaches, and preaches . . . I held and believed that the sun is the center of the universe and is immovable, and that the earth is not the center and is movable; willing, therefore, to remove from the minds of your Eminences, and of every Catholic Christian, this vehement suspicion [of heresy] rightfully entertained against me, . . . I abjure, curse and detest the said errors and heresies, . . . and I swear that I will never more in future say or assert anything verbally, or in writing, which may give rise to a similar suspicion of me . . . But if it shall happen that I violate any of my said promises, oaths and protestations (which God avert!), I subject myself to all the pain and punishments which have been decreed . . . against delinquents of this description.[42]

Thereupon he was sentenced to life imprisonment in a Roman dungeon. The sentence was later commuted and he died, a blind and broken man, in 1642. But not even in death did the pious inquisitors leave him in peace. They destroyed many of his manuscripts, disputed his right to burial in consecrated ground, and denied him a monument in the ludicrous hope that this brilliant thinker and his work would be forgotten. 250 years later, Sir Oliver Lodge commented:

> Poor schemers! Before the year was out, an infant was born in Lincolnshire, whose destiny it was to round and complete and carry forward the work of their victim, so that, until man shall cease from the planet, neither the work nor its author shall have need of a monument.

SUCH was the ugly face of the Middle Ages. It is not surprising that mathematics made little progress; toward the Renaissance, European mathematics reached a level that, roughly, the Babylonians had attained some 2,000 years earlier[43] and much of the progress made was due to the knowledge that filtered in from the Arabs, the Moors and other Muslim peoples, who themselves were in contact with the Hindus, and they, in turn, with the Far East.

The history of π in the Middle Ages bears this out. No significant progress in the method of determining π was made until Viète discovered an infinite product of square roots in 1593, and what little progress there was in the calculation of its numerical value, by various modifications of the Archimedean method, was due to the decimal notation which began to infiltrate from the East through the Muslims in the 12th century.

Arab mathematics came to Europe through the trade in the Mediterranean, mainly via Italy; ironically, the other stream of mathematics was the Church itself.[44] Not only because the mediaeval priests had a near monopoly of learning, but also because they needed mathematics and astronomy as custodians of the calendar. Like the Soviet High Priests who publish *Pravda* for others but read summaries of the *New York Times* themselves, so the mediaeval Church condemned mathematics as devilish for others, but dabbled quite a lot in it itself. Gerbert d'Aurillac, who ruled as Pope Sylvester II from 999 to 1003, was quite a mathematician; so was Cardinal Nicolaus Cusanus (1401-1464); and much of the work done on π was done behind thick cloister walls. And just like the Soviets did not hesitate to spy on the atomic secrets of bourgeois pseudo-science, so the mediaeval Church did not hesitate to spy on the mathematics of the Muslim infidels.

GALILEO GALILEI
(1564-1642)

Discovered the law of the pendulum; radically improved the telescope; discovered the satellites of Jupiter, sunspots, the rotation of the Sun, and the libration of the Moon; proved the uniform acceleration of all bodies falling to earth; doubted the infinite velocity of light and suggested how to measure its velocity; established the concept of relative velocity. Above all, he discovered the laws of motion, though he was not yet able to formulate them quantitatively.

Adelard of Bath (ca.1075-1160) disguised himself as a Muslim and studied at Cordoba;[45] he translated Euclid's *Elements* from the Arabic translation into Latin, and Ptolemy's *Almagest* from Greek into Latin. When Alfonso VI of Castile captured Toledo from the Moors in 1085, he did not burn their libraries, containing a wealth of Muslim manuscripts. Under the encouragement of the Archbishop of Toledo, a veritable intelligence evaluation center was set up. A large number of translators, the best known of whom was Gerard of Cremona (1114-1187), translated from Arabic, Greek and Hebrew into Latin, at last acquainting Europe not only with classical Greek mathematics, but also with contemporary Arab algebra, trigonometry and astronomy. Before the Toledo leak opened, mediaeval Europe did not have a mathematician who was not a Moor, Greek or a Jew.[46]

One of the significant mediaeval European mathematicians was Leonardo of Pisa (ca. 1180-1250), better known by his nickname Fibonacci ("son of Bonaccio"). Significantly, he was an Italian merchant, so that he worked within one of the Arabic infiltration routes. In 1202, he published a textbook using algebra and the (present) Hindu-Arabic numerals. He is best known for the Fibonacci sequence

$$1, 1, 2, 3, 5, 8, 13, 21, 34, 55, 89, \ldots\ldots,$$

where each number (after the initial 1's) is the sum of the two preceding ones. This sequence turns up in the most surprising places (see figure on p. 24), and its applications range from the growth of pineapple cells to the heredity effects of brother-sister incest. There is a *Fibonacci Quarterly* in the United States in our own day, for the fertility of this sequence appears to be endless.

But Fibonacci also worked on π, though his progress on this point was not as impressive as his other achievements. Like Archimedes, Leonardo used a 96-sided polygon, but he had the advantage of calculating the corresponding square roots by the new decimal arithmetic. Theoretically, his work (published in *Practica geometriae*, 1220) is not as good as Archimedes', whose approximations to the square roots are always slightly on the low side for the circumscribed polygon, and slightly on the high side for the inscribed polygon; since the square roots appeared in the denominator, this assured Archimedes of getting the correct bounds. Fibonacci used no such care, extracting the square roots as nearly as he could, but he was lucky: His bounds turned out to be more accurate than Archimedes', and the mean value between them,

$$\pi = 864 : 275 = 3.141818$$

is correct to three decimal places.

Otherwise there was little progress. Gerbert (Pope Sylvester) used the Archimedean value $\pi = 22/7$, but in the next 400 years we also find the Babylonian 3 1/8 and the Egyptian $\pi = (16/9)^2$ in various cloister correspondence.[47] The documents of the time mostly show the low level of mediaeval mathematics. Franco von Lüttich, for example, wrote a long treatise on the squaring of the circle (ca. 1040), which shows that he did not even know how to square a rectangle[47] (p. 51), and Albert von Sachsen (died 1390) wrote a long treatise *De quadratura circuli* consisting for the most part of philosophical polemics. The crux of the problem is brushed away by saying that "following the statement of many philosophers," the ratio of circumference to diameter is exactly 22/7; of this, he says, there is proof, but a very difficult one.[47] Dominicus Parisiensis, author of *Practica geometriae* (1346), distinguishes himself above his contemporaries by at least knowing that $\pi = 22/7$ was an approximation. So did the Viennese Georg Peurbach (1423-1461), who knew Greek and some of the history of π; he understood Archimedes' derivation of $\pi > 22/7$, and knew both the Ptolemeian $\pi = 377 : 120$, and the Indian value $\pi = \sqrt{10}$.

Perhaps the only interesting contribution of this sorry time is that of Cardinal Nicolaus Cusanus (1401-1464), a German who worked in Rome from 1448 until his death. Although his work on π was not very successful and his approaches were more ecclesiastic than mathematical,[48] he did find a good approximation for the length of a circular arc. His derivation is somewhat tiresome,[49] but in modern terminology, it amounts to

$$\operatorname{arc}\theta \approx \frac{3\sin\theta}{2+\cos\theta} \tag{1}$$

To find the quality of this approximation, we multiply (1) by $(2 + \cos\theta)$ and expand both sides in series (even though Brook Taylor is still two and a half centuries away), obtaining

$$3\theta - \theta^3/2! + \theta^5/4! - \ldots \approx 3\theta - \theta^3/2! + 3\theta^5/5! - \ldots$$

so that up to the third-order term the two expressions are identical, and the fifth order term differs only by 3/5. The approximation is therefore excellent for

$$\theta^2 \ll 12;$$

for $\theta = 36°$, the error is about 2 minutes of arc. Well done, Cardinal.

It is doubtful whether some two centuries later, the great Dutch physicist Christiaan Huygens (1629-1693) was familiar with Cusanus' work, but in his treatise *De circuli magnitude inventa*, which we shall meet again in Chapter 11, he derived a theorem suggesting a construction for the rectification of a circular arc as shown in the figure below, and which makes arc $AB \approx AB'$. The construction was first suggested by Snellius, and it is equivalent to the Cardinal's approximation: We have

$$AB' = 3r\tan\beta \tag{2}$$

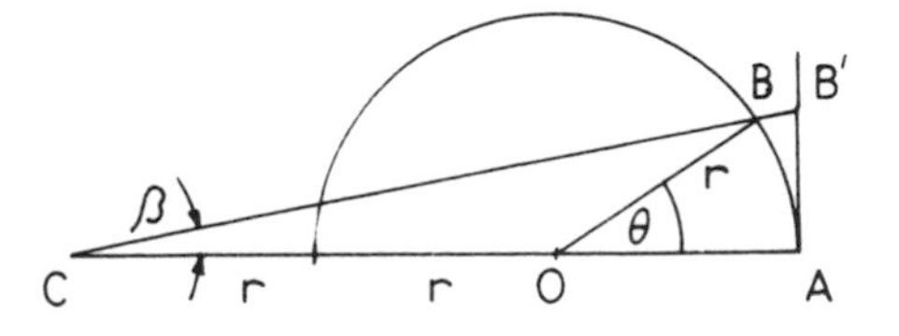

Approximate rectification of an arc suggested by one of Huygens' theorems, and equivalent to Cardinal Cusanus' approximation: $AB' \approx AB$.

and from the triangle *OCB*,

$$\sin\beta = \frac{\sin\theta}{\sqrt{5 + 2\cos\theta}} \tag{3}$$

From (3) we find $\tan\beta$, substitute in (2), and obtain after some trigonometric manipulation

$$AB' = \frac{3r\sin\theta}{2 + \cos\theta} \tag{4}$$

If this is set equal to arc $AB = r\theta$, it will be seen that Huygens' construction is equivalent to Cusanus' approximation (1).

But Huygens already lived in happier times.

9

AWAKENING

E pur si muove!
(And nonetheless it moves!)
Giordano Bruno's last cry from the burning stake, 16th February, 1600.[50]

THE classification of history into ancient, mediaeval and modern is but a symptom of the white man's arrogance; for while Europe was suffering the imbecility of the Dark Ages, the rest of the world went on living. Yet in the struggles of oriental science we find the same type of patterns and tendencies as in Europe, though not necessarily at corresponding times. Just as the torch of Alexandria was extinguished by militaristic Rome, so the intellectual life of Babylon was wiped out by the militaristic Assyrians, and the golden age of Muslim science was stifled by the militaristic Turks; in India and China the story is similar. And again, we find the confrontations between religion and science. In the vein of Tertullian, the Muslim philosopher al-Ghazzali (1058-1111) wrote that scientific studies shake men's faith in God and undermine religion, and that they lead to loss of belief in the origin of the world and the Creator;[51] and several Muslim fanatics, like their Christian counterparts, proved their piety by burning libraries. In China, too, the emperor Tsin Shi Hwang-di (3rd century B.C.) was warned by his advisors of the idling scholars whose influence was founded on books; and he ordered all literature in China burned, excepting only books on medicine, husbandry and divination, and those in the hands of the seventy official scholars. And again, we find the quaint mixture of mathematics and mumbo-jumbo that can be found in the mysticism of the Pythagoreans. For example, the able Chinese mathematician Sun-Tsu (probably 1st century A.D.) wrote a textbook containing, for the most part, sound mathematics; but it also contains this problem:

> A pregnant woman, who is 29 years of age, is expected to give birth to a child in the 9th month of the year. Which shall be her child, a son or a daughter?

The solution is truly Pythagorean:

> Take 49; add the month of her child-bearing; subtract her age. From what remains, subtract the heaven 1, subtract the earth 2, subtract the man 3, subtract the four seasons 4, subtract the five elements 5, subtract the six laws 6, subtract the seven stars 7, subtract the eight winds 8, subtract the nine provinces 9. If the remainder be odd, the child shall be a son; and if even, a daughter.

Note that the remainder is negative, so that a Chinese student of the 1st century would be stunned, because he could scarcely know the concept of a negative number, let alone its parity. To say very impressively nothing at all is the secret of all such oracles; it is also the secret of computerized astrology and the trumpets of Madison Avenue.

AS we have seen, European mathematics was far behind that of the Muslim world. Yet as the Middle Ages ended, Europe was catching up, and by the 17th century it was so far ahead that the rest of the world never caught up with it again, and to Europe we now return.

The end of the Middle Ages was heralded by three events, all of which took part in the latter half of the 15th century.

First, in 1453 the Turks captured Constantinople. The significance of this event is not that they grabbed and sacked a city ruled by those who had grabbed and sacked it before them, but that they had breached its walls with guns. The era of gun powder mean the end of the feudal fiefs who often defied their kings behind the walls of their castles. The local tyranny of the feudal aristocracy was replaced by the centralized tyranny of the divine kings and emperors. And the absolute power of the Church was broken; where the Church had patron-

Christopher Columbus
(1451–1506)

Vasco da Gama
(1469–1525)

Ferdinand Magellan
(1480–1521)

ized the nominal ruler, the central and powerful ruler was now patronizing the Church.

Second, the turn of the fifteenth century saw the great discoveries of the rest of the world. Christopher Columbus discovered America; Vasco da Gama rounded the Cape of Good Hope (1499) on his voyage to India; and Ferdinand Magellan's flotilla circumnavigated the globe (1519-1522). The far-reaching consequences of these events included the need for better clocks, better astronomy, better trigonometry, and the resulting stimulation of the exact sciences. It is probably not accidental that the mathematicians appearing on the scene come increasingly from England, France and Holland, countries that suddenly found themselves in the center of the maritime world, and where (unlike Spain and Portugal) the influence of the Church was rapidly waning.

Third, and perhaps most important, the 15th century saw the revolution in the dissemination of information brought about by printing from movable and reusable type. This had been used in China since the 9th century (mostly for printing charms), but in the 15th century several Europeans started experimenting with it, and though no single man is usually responsible for an epoch-making invention of this kind, credit is traditionally given to Johannes Gutenberg of Mainz, who printed an edition of the Bible by this method in 1456. Until then, manuscripts were usually copied by hand, as a rule by monks in monasteries, one of them dictating from a master copy to an assembly of scribes, or by the painstaking process of carving a wooden block for an entire page to be printed.

More than anything else, the invention of the letterpress broke the Church's monopoly of learning. Henceforth, books were available to ever wider sections of the population, and the subversive little invention eventually tumbled the thrones of Europe's tyrants. (No tyrants were ever more acutely aware of this than the Jenghis Khans in the Kremlin: In the USSR, not only presses, but even little mimeographs are jealously guarded by a multiple net of security regulations such as civilized countries do not even use for presses printing currency.) Like other books, mathematical textbooks could now be mass produced. It was a sign of the times that mathematical textbooks were no longer published only in Latin, the language of the educated, but also in the national languages. Advanced mathematics continued to be written in Latin until the beginning of the 19th century, but textbooks of elementary mathematics began to appear in English, French, German, Italian, Dutch and other languages (some of them even before the invention of the letterpress).

The invention of the letterpress affected mathematical works in yet another way. In the late Middle Ages, mathematicians would meet in challenge disputes, often for considerable sums of money, in which mathematicians posed problems to each other. University positions were subject to renewal, and the university authorities would be influenced by the outcome of these contests. In 1535, for example, a contest between Antonio Fiore and Niccolo Tartaglia in Venice involved 30 banquets that the loser was to give for the winner and his friends. Naturally, these mathematical prizefighters jealously guarded the secrets of their trade. The contest above, for example, centered around the equation

$$x^3 + ax = b;$$

Tartaglia knew how to solve it and confided the secret to the colorful founder of probability theory Gerolamo Cardano (1501-1576), but swore him to secrecy, and there was much intrigue, polemics and unpleasantness when Cardano, who later found the solution of the general cubic equation, published his results, even though he gave credit to Tartaglia (who had himself obtained the trick through an intermediary from Scipione del Ferro of Bologna, the latter having discovered it some time between 1500 and 1515).[52] The letterpress changed all this; for mathematicians found a surer road to renown than prizefighting: Publish or perish was the name of the new game.

THE story of π during this age of the Renaissance was mainly one of ever more accurate numerical values of the constant; the theory was

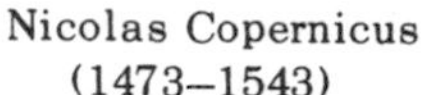

Nicolas Copernicus (1473–1543) | Tycho Brahe (1546–1601) | Johann Kepler (1571–1630)

still essentially based on the Archimedean polygons. Apart from the Hindu-Arabic numerals and decimal fractions that had filtered from the Arabs into Europe during the Middle Ages, there were now two new instruments for numerical calculations at the disposal of the circle squareres: trigonometric functions and logarithms.

Trigonometric functions (in one form or another) had already been used by Ptolemy and the other scientists at the University of Alexandria, but accurate tables now became available through the works of the Renaissance astronomers, foremost among whom were Nicolas Copernicus (1473-1543) and Johann Kepler (1571-1630). Both these trailblaizing astronomers are connected with tables of trigonometric functions. Copernicus introduced the secant into trigonometry and was apparently the first to calculate a table; Kepler popularized, refined and explained Napier's logarithms of trigonometric functions. From 1600 to 1612 Kepler worked at the Royal Castle of Prague for the king of Bohemia and emperor of Austria Rudolf II of Hapsburg, where he succeeded the Danish astronomer Tycho Brahe. Rudolf II surrounded himself with astrologers and alchemists, whom he hired to find the Stone of the Wise, by whose magic one could make gold. Kepler was not the last scientist hired to produce gold, and who produced something of more lasting value instead; the tradition continues among the scientists of our day.

Logarithms were discovered at the beginning of the 17th century indpendently by Jobst Bürgi, a Swiss clockmaker working in Prague, and John Napier, a Scottish nobleman and amateur mathematician, whose many activities included the publication of *The Plaine Discouery of the whole Reuelation of Saint John* (in which he argued that the Pope was the Anti-Christ), contracting for the discovery of treasure, devising warlike machines for the defense of Britain against

Philip of Spain, promoting astrology, and recommending salt as a fertilizer. Napier's book *Mirifici logarithmorum canonis descriptio* (A Description of the Wonderful Rule of Logarithms) was published earlier (1614) than Bürgi's (1620), and he is generally regarded as the discoverer of logarithms.

These developments, naturally, facilitated the calculation of π by the Archimedean method or its modifications, and accuracies far beyond any possible practical use were obtained. We shall come back to this point in the next chapter, but first we shall examine the theoretical progress made during this period. The main achievement was that of another amateur mathematician, François Viète, Seigneur de la Bigotière (1540-1603). He was a lawyer by profession and rose to the position of councillor of the *Parlement* of Brittany, until forced to flee during the persecution of the Huguenots. The next six years or so during which he was out of favor, he spent largely on mathematics. With the accession of Henry IV, a former Huguenot, Viète was restored to office, becoming Master of Requests (1580) and a Royal Privy Councillor (1589). He endeared himself to the king by breaking the Spanish code made up of some 500 cyphers, thus enabling the French to read all secret enemy dispatches. Thereupon the Spanish, with singular one-track-mindedness, accused him of being in league with the devil.

Viète made important contributions to arithmetic, algebra, trigonometry and geometry. He also introduced a number of new words into mathematical terminology, some of which, such as *negative* and *coefficient*, have survived. His attack on π, though still proceeding along general Archimedean lines, started with a square rather than a hexagon, and resulted in the first analytical expression giving π as an infinite sequence of algebraic operations.

His procedure consisted (essentially) of relating the area of an n-sided polygon to that of a $2n$-sided polygon (see figure on opposite page). The area of an n-sided polygon is

$$A(n) = n \text{ times area of triangle } OAB$$

$$= \tfrac{1}{2} n r^2 \sin 2\beta \quad (1)$$

$$= n r^2 \cos\beta \sin\beta \quad (2)$$

Similarly,

$$A(2n) = n r^2 \sin\beta \quad (3)$$

so that from (2) and (3),

$$A(n)/A(2n) = \cos\beta \quad (4)$$

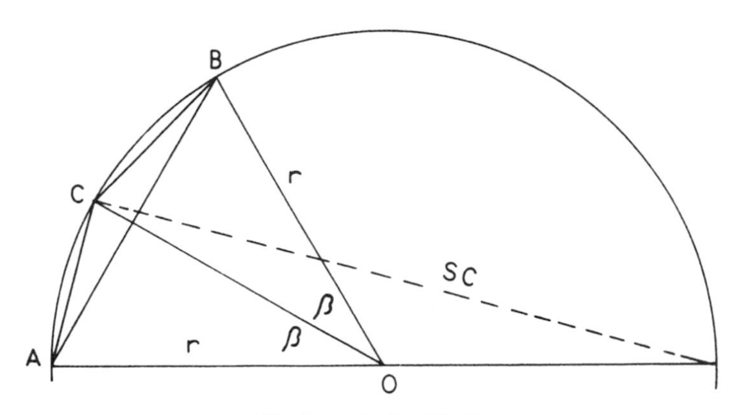

Viète's method of finding π.

and if we double the sides of the polygon again, we have

$$\frac{A(n)}{A(4n)} = \frac{A(n)}{A(2n)} \times \frac{A(2n)}{A(2^2 n)} = \cos\beta \cos(\beta/2) \qquad (5)$$

Continuing like this k times, we obtain

$$\frac{A(n)}{A(2^k n)} = \frac{A(n)}{A(2n)} \times \frac{A(2n)}{A(4n)} \times \ldots \times \frac{A(2^{k-1} n)}{A(2^k n)}$$

$$= \cos\beta \cos(\beta/2) \ldots \cos(\beta/2^k) \qquad (6)$$

But if k tends to infinity, the area of a regular polygon with 2^k sides is indistinguishable from that of a circle, so that

$$\lim_{k \to \infty} A(2^k n) = \pi r^2. \qquad (7)$$

Substituting (7) and (2) in (6), we have

$$\pi = \frac{\frac{1}{2} n \sin 2\beta}{\cos\beta \cos(\beta/2) \cos(\beta/2^2) \cos(\beta/2^3) \ldots} \qquad (8)$$

Viète chose a square to start with, so that $n = 4$, $\beta = 45°$, $\cos\beta = \sin\beta = \sqrt{\frac{1}{2}}$.

Also, each of the cosine factors in (8) is expressible in terms of the preceding factor through the half-angle formula

$$\cos(\theta/2) = \sqrt{(\tfrac{1}{2} + \tfrac{1}{2}\cos\theta)} \qquad (9)$$

so that we finally have

$$\pi = \frac{2}{\sqrt{\tfrac{1}{2}} \times \sqrt{(\tfrac{1}{2} + \tfrac{1}{2}\sqrt{\tfrac{1}{2}})} \times \sqrt{[\tfrac{1}{2} + \tfrac{1}{2}\sqrt{(\tfrac{1}{2} + \tfrac{1}{2}\sqrt{\tfrac{1}{2}})}]} \times \ldots} \qquad (10)$$

and this is Viète's expression, published in 1593 in his *Variorum de rebus mathematicis responsorum liber VIII* (Various mathematical problems, vol. 8). Actually, his derivation used the supplementary chord of a polygon, that is, the chord joining a point of a polygon on a circle to the other end of the diameter, marked *SC* in the figure on p. 93. (Note that this is the supplementary chord of a $2n$-sided, not of an n-sided polygon.) Viète showed that the supplementary chord of n-sided polygon is to the diameter as the area of the polygon with n sides is to that of the polygon with $2n$ sides. But since that ratio is simply $\cos\beta$, the derivation given above conveys the essence of Viète's procedure.

Viète's result represents one of the milestones in the history of π, and it is also the high point of renaissance mathematics connected with π; therefore it deserves a few comments.

First, we note that Viète was still imprisoned by the idea of the Archimedean polygon; he is, in fact, one of the last men to use a polygon in furthering the theory of calculating π (the others were Descartes, Snellius and Huygens, as we shall see in Chapter 11), though he was far from last in using it for numerical evaluation.

Second, Viète was the first in history to represent π by an analytical expression of an infinite sequence of algebraic operations. (As a matter of fact, he was also the first to use the term "analytical" in mathematical terminology, and the term has survived.) The idea of continuing certain operations *ad infinitum* was of course much older. Archimedes had used it, and before him Antiphon expressed the principle of exhaustion, as we have seen on p. 37. Viète was familiar with the Greek classics, and he refers to Antiphon in his treatment; his approach executed Antiphon's idea mathematically. However, there is a vast difference between expressing an idea qualitatively and giving quantitative instructions how to execute it, and Viète was the first to achieve this. Viète's expression, in fact, is the first known use of an infinite product, whether connected with π or not.

Third, Viète was a typical child of the Renaissance in that he freeley mixed the methods of classical Greek geometry with the new Arabic art of algebra and trigonometry. The idea of substitution is algebraic, and the square roots in his expression come from the trigonometric half-angle formula for the cosine, but otherwise his treatment is entirely Greek, based on considerations of area ratios involving the supplementary chord. Had he tried to express the mouthful *ratio of the supplementary chord to the diameter* trigonometrically, he would have found that it equals cos β, a much more easily manipulated quantity. He would then have obtained our formula (8) above, and a man of Viète's stature could hardly have overlooked that by expressing β in radian measure and setting $\theta = 2\pi/n$, he would have obtained the formula

$$\theta = \frac{\sin\theta}{\cos(\theta/2)\cos(\theta/2^2)\cos(\theta/2^3)\ldots}, \qquad (\theta < \pi) \qquad (11)$$

which Leonard Euler obtained in a quite different way almost 200 years later. Viète's result (10) is a special case of (11) for $\theta = \pi/2$.

Fourth, Viète did not yet know the concept of convergence and did not worry whether his infinite sequence of operations would "blow up" or not. We need not worry either, because as far as the formula for π is concerned, it is sufficient to take k arbitrarily large, but finite. However, if you are a friend of mathematical rigor and this kind of "sloppy engineering mathematics" disgusts you, rest assured: The convergence of Viète's formula was proved by F. Rudio in 1891.[53]

Fifth, it should be noted that Viète's formula is of almost no use for numerical calculations of π; the square roots are much too cumbersome, and the convergence is slow. Viète himself did not use it for his calculation correct to 9 decimal places; he used the Archimedean method without substantial modification by taking a polygon of 393,216 sides (the number is obtained by 16 successive doublings of the original hexagon). This enabled him to reduce the Archimedean bounds to

$$3.1415926535 < \pi < 3.1415926537 \qquad (12)$$

but this is only a minor success compared with the formula (10) which he derived. In the vast majority of practical applications, $\pi = 22/7$ is good enough, and to obtain better and better approximations is only a matter of drudgery. In contrast, Viète's formula (10) is an entirely new formulation, one that can be manipulated and investigated. Indeed, if no better expressions had been discovered since then, this one could have served for investigating some of the properties of π that no number of decimal places can reveal.

Finally, the fact that Viète introduced a number of mathematical terms that have survived to the present is a point of interest. Terminology and symbolism cannot, by themselves, solve anything; but if not conveniently chosen, they can spoil a lot. The long sentence for the circle ratio (p. 76) could not be squared or subjected to other mathematical operations as the symbol π can, and Viète's example shows that one can miss an important turning to a wide new field by using Greek geometry instead of the concise symbolism of trigonometry. In Viète's time, mathematical notation was still a long way from what it is now. The operator symbols $+$, $-$, $=$ had only very recently been introduced, and algebra often described an equation in words rather than symbols. The unknown quantity (our "x") had a strange name. The Italians, through whom algebra mostly came to Europe from the Arabs, called it *cosa*, the "thing" (the same word as the Mafia uses in *cosa nostra*), and the word went into other languages, e.g., in German it became *die Coss*, and in Latin it became *numerus cossicus*; in English, algebra was known as *the Cossike arte*. Robert Recorde (1510-1558) tells us that *in nombers Cossike, all nombers haue not rootes; but soche only emongest simple cossike nombers are rooted, whose nomber hath a roote, agreable to the figure of his denomination.*

There are other quaint little tid-bits in mathematical terminology. For example, the origin of the words *tangent* and *secant* are clear enough, but where does *sine* come from? It comes from a translator's error in the Toledo translation center (p. 83). Arabic script, like Hebrew script, consists of consonants, with the vowels punctuated underneath, and the latter are often omitted. The sine, which one would expect to be called "half-chord" in analogy with the secant and tangent, was given a name by the Hindus, which the Arabs took over, and which they spelled by the consonants *jb*. When Robert of Chester, one of the Toledo translators, translated al-Khowarizmi's *Algebra* from Arabic into Latin in 1145, he encountered this word without knowing its Hindu origin; supplying the missing vowels, he found the Arabic word for bay or inlet, and the Latin for bay, inlet or cavity is *sinus*.

But back to π. The Renaissance was a golden age of the amateur mathematician. Viète was Royal Privy Councillor, Bürgi a clockmaker, Napier a landowner and theologist, Cardano a physician. Two more amateur mathematicians of the Renaissance deserve mention. Their mathematical achievements were not very impressive, but they were great artists: Albrecht Dürer (1471-1528) and Leonardo da Vinci (1452-1519). Dürer studied geometry, or what today would more closely be described as descriptive geometry, because as an artist he was interested in perspective and the proportions of the human form. Leonardo da Vinci was interested in mathematics for the simple reason that he was interested in everything; or at least so one would judge on reading the copious material which he entrusted to his notebooks and other manuscripts. Dürer, in his *Underweysung der messung mit dem zirckel und richtsheyt* (Instruction for mensuration with compasses and straightedge), 1525, uses the Babylonian value $\pi = 3\ 1/8$. Leonardo's voluminous manuscripts contain at least two entries directly connected with the squaring of the circle. One is the construction by rearrangement of circular segments which has already been discussed (p. 18-19); the other is an original idea of how to square the circle: Take a wheel (cylinder) whose thickness (height) equals half the radius of the wheel; roll it one revolution, and the area of the rectangular track it leaves is equal to the area of the circle of the wheel (base of the cylinder). Leonardo's words are *La intera revolutione della rota quala la grossezza sia equale al suo semidiamitro lasscia di se vesstigio equale alla quadratura del suo cierchio.* The "semidiamitro" is evidently an oversight, for half the radius, not half the diameter, must be used; in that case the track left by the wheel is indeed $2\pi r \times \frac{1}{2} r = \pi r^2$.

Viète's result of finding π accurately to 9 decimal digits was preceded and followed by similar calculations, all based on the Archimedean polygons; they did not bring anything significantly new, and we delay the discussion of these indefatigable digit hunters to the next chapter; in the meantime we pause to note some interesting oddities of the history of π in the Renaissance.

There is, for example, the earliest of my several competitors, a French scholar by the name of Johannes Buteo (1492-1572); in 1559 he published a book *De quadratura circuli*, which seems to be the first book that amounts to a history of π and related problems. Buteo, unlike many of his contemporaries, was thoroughly familiar with the Archimedean method, and gives a survey of the methods used in antiquity and the Middle Ages.

Then there is Joseph Scaliger (1540-1609; not to be confused with Julius Scaliger, 1484-1558, who first published a book abusing and ridiculing Gerolamo Cardano, and mistakenly believing it had caused him to die, followed it by a grandiose and heartrending eulogy, heaping lavish praise on the supposed victim[54]). Joseph Scaliger was a famous philologist at the University of Leyden, who in 1594 published a work *Cyclometria elementa* in which he purports to demonstrate that already the perimeter of a 12-sided polygon is greater than the circumference of the circumscribed circle, so that it is to no purpose to use polygons with more than 12 sides for the calculation of π. It was possible, maintained Scaliger, for something to be geometrically true, but arithmetically false.

It is to be hoped that Scaliger was a better philologist than mathematician.

Unlike Scaliger, Michael Stifel (ca. 1487-1567) was a significant mathematician, the most important German algebraist of the 16th century. He was a former monk, turned itinerant Lutheran preacher, and for a time professor of mathematics at the University of Jena. Although he was mainly interested in algebra, he also dabbled in circle squaring, but came to the conclusion that attempts to square the circle were futile. Although he was right, his opinion, at the time, could have been little more than a lucky guess, probably with an admixture of sour grapes.

10

The Digit Hunters

So, Nat'ralists observe, a Flea
Hath smaller Fleas that on him prey.
And these have smaller Fleas to bite'em,
And so proceed ad infinitum.
JONATHAN SWIFT
(1667-1745)

IT has already been pointed out that the invention of decimal fractions and logarithms greatly facilitated numerical calculations in the late 1500's and the early 1600's, and this is reflected in the history of π, for about this time people started to calculate its value to an ever increasing number of decimal places, each new digit increasing the accuracy of the former approximation by no less than 10 times. The process continued beyond any possible practical use of so many decimal places; by the end of the 16th century, π was known to 30 decimal places, by the end of the 18th century it was known to 140 places, by the end of the 19th century it had been calculated to 707 places (though later only 526 of them proved to be correct), and the digital computer of the 20th century has raised this number to a whopping 500,000 — perhaps more by the time you read this (see Chapter 18).

Archimedes calculated π to the equivalent of two decimal places, and at first the hunt for greater accuracy may have been dictated by practical needs. Later, especially after the advent of the differential calculus and infinite series, the number of decimal places may have been used to demonstrate the quality of the method of calculation. Perhaps some investigators hoped to discover a periodicity in the ever lengthening sequence of digits. Had this been so, they would have been able to express π as the ratio of two integers, for if a certain sequence of d decimal digits kept recurring, then the fractional part would be the sum of the geometric series

$$S = a_0(q + q^2 + q^3 + \dots), \qquad q = 10^{-d},$$

where a_0 is the number formed by the first d digits, so that by summing this geometric series, one would obtain

$$\pi = 3 + \frac{a_0}{1 - 10^{-d}}$$

However, as was shown by Lambert in 1767, π is not a rational number, i.e., it cannot be expressed as a ratio of two integers, and this showed hopes of periodicity futile. Another possible reason may have been that the calculation of π presented a challenge to find better methods of numerical analysis, for numerical calculations are far from easy when the number of operations becomes large and every little trick could save hours of computations.

But for the most part, I suspect, the driving force behind these calculations was the spirit that makes people go over the Niagara Falls in a barrel or to top the world record of pole sitting by another 20 minutes. The digits beyond the first few decimal places are of no practical or scientific value. Four decimal places are sufficient for the design of the finest engines; ten decimal places would be sufficient to obtain the circumference of the earth within a fraction of an inch if the earth were a smooth sphere (in proportion, it is smoother than a billiard ball). The most exacting requirement that I can think of in a practical application is a not very common case of computer programming. Computers do the actual arithmetic operations only with rational numbers, which are used to approximate irrational numbers. In rounding off the last significant figure, they can sometimes play quite treacherous tricks on the programmer, and to guard against the effect of the rounding error in computations that involve long sequences of certain operations (for example, very small differences between very large numbers), special precautions must be taken. In FORTRAN, a widely used computer language, this is achieved by a command called DOUBLE PRECISION. This will result in certain operations being carried out with at least double the usual number of significant figures. The actual number of decimal places varies from computer to computer, but usually as many as 17 decimal places can be stored for a double precision constant. For this extreme case, the corresponding value of π is

$$\pi = 3.14159\ 26535\ 89793\ 24,$$

where the last digit results from rounding off 38. Or we let the computer tell us: We ask it for 4 arctan 1 (which is π) with double precision by entering the FORTRAN statement

```
PI = 4.0 * DATAN(1.0 D 0)
```

The Double Precision routine is used only when really necessary, because it can be more trouble than it is worth, and even then 17 decimal places are usually more than enough.

There is no practical or scientific value in knowing more than the 17 decimal places used in the foregoing, already somewhat artificial, application. In 1889, Hermann Schubert, a Hamburg mathematics professor, made this point in the following consideration.[56]

> Conceive a sphere constructed with the earth at its center, and imagine its surface to pass through Sirius, whis is 8.8 light years distant from the earth [that is, light, traveling at a velocity of 186,000 miles per second, takes 8.8 years to cover this distance]. Then imagine this enormous sphere to be so packed with microbes that in every cubic millimeter millions of millions of these diminuitive animalcula are present. Now conceive these microbes to be unpacked and so distributed singly along a straight line that every two microbes are as far distant from each other as Sirius from us, 8.8 light years. Conceive the long line thus fixed by all the microbes as the diameter of a circle, and imagine its circumference to be calculated by multiplying its diameter by π to 100 decimal places. Then, in the case of a circle of this enormous magnitude even, the circumference so calculated would not vary from the real circumference by a millionth part of a millimeter.
>
> This example will suffice to show that the calculation of π to 100 or 500 decimal places is wholly useless.

But microbes or no microbes, the digit hunters plodded ahead through the centuries.

The Dutch mathematician and fortification engineer Adriaan Anthoniszoon (1527-1607) found the value 355/113, which is correct to 6 decimal places.

This record was broken by François Viète in 1593 by the value given on p. 95, which is correct to 9 decimal places.

But already in the same year his record fell to Adriaen van Rooman (1561-1615), a Dutchman who used the Archimedean polygons with 2^{30} sides, and calculated 15 decimal places.

Three years later his record was broken by another Dutchman, Ludolph van Ceulen (1539-1610), professor of mathematics and military science at the University of Leyden. In his paper *Van den Circkel* (1596), he reports using a polygon with 60×2^{29} sides, which yielded the value of π to 20 decimal places. The paper ends with "Whoever wants to, can come closer." But nobody wanted to, except Ludolph himself. In *De Arithmetische en Geometrische fondamenten*, published posthumously by his wife in 1615, he gives π correct to 32 places, and according to Snell's report in 1621, he topped this later by three more places. Tropfke[57] states that these last three digits are engraved on his tombstone in the Peter Church at Leyden; this seems to invalidate vague references by other historians, according to which all 35 digits were engraved in his tombstone, but that the stone has been lost. In any case, Ludolph's digit hunting so impressed the Germans that to this day they call π *die Ludolphsche Zahl* (the Ludolphine number).

But then came the boys with the big guns. The differential calculus was discovered in the 17th century, and with it a multitude of infinite series and continued fractions for π, as we shall see in coming chapters. The astronomer Abraham Sharp (1651-1742) used an arcsine series to obtain 72 decimal digits, and shortly afterwards, in 1706, John Machin (1680-1752) used the difference between two arctangents to find 100 decimal places; but he was beaten by the French mathematician De Lagny (1660-1734), who piled another 27 digits on top of this result in 1717. This record of 127 digits seems to have stood until 1794, when Vega (1754-1802), using a new series for the arctangent discovered by Euler, calculated 140 decimal places. Vega's result showed that De Lagny's string of digits had a 7 instead of an 8 in the 113th decimal place.

It should be noted that during the 18th century Chinese and Japanese digit hunters were at work also, although the Japanese records lagged behind the European ones, just as the Chinese records were retarded with respect to the Japanese. The Japanese mathematician Takebe, using a 1,024-sided polygon, found π to 41 decimal places in 1722, and Matsanuga, using a series, found 50 decimal places in 1739. Thereafter, the Japanese seem to have had more sense than their European colleagues; they continued to study series yielding π, but wasted no more time on digit hunting.

The Viennese mathematician L.K. Schulz von Strassnitzky (1803-1852) used an arctangent formula to program the forerunner of the computer, a calculating prodigy. This one was Johann Martin Zacharias Dase (1824-1861), about whom we shall have more to say below. In 1844 he calculated π correct to 200 places in less than two months. They are reproduced below:

$$
\begin{aligned}
\pi = 3.&14159\ 26535\ 89793\ 23846\ 26433\\
&83279\ 50288\ 41971\ 69399\ 37510\\
&58209\ 74944\ 59230\ 78164\ 06286\\
&20899\ 86280\ 34825\ 34211\ 70679\\
&82148\ 08651\ 32823\ 06647\ 09384\\
&46095\ 50582\ 23172\ 53594\ 08128\\
&48111\ 74502\ 84102\ 70193\ 85211\\
&05559\ 64462\ 29489\ 54930\ 38196.
\end{aligned}
$$

Before Dase, William Rutherford had calculated 208 decimal places in 1824; but from the 153rd decimal place they disagreed with Dase's figures, and the discrepancy was resolved in Dase's favor when Thomas Clausen (1801-1885) published 248 decimal places in 1847.

And still the craze went on. In 1853, Rutherford gave 440 decimal places, and in 1855 Richter calculated 500 decimal places. When William Shanks published 707 places in 1873-74 (*Proceedings of the Royal Society*, London), he probably thought that his record would stand for a long time. And so it did. But in 1945, Ferguson found an error in Shanks' calculations from the 527th place onward,[93] and in 1946, he published 620 places; using a desk calculator, he subsequently found 710 places in January 1947, and 808 places in September of the same year.[93] This was the record that succumbed to the computer in 1949, and we shall return to the story in Chapter 18.

To appreciate the stupendous work necessary to achieve such results, it is necessary to recall that a computer can perform an arithmetical operation, such as adding or multiplying two numbers, in less than one millionth of a second; but even a contemporary computer takes about 40 seconds (not counting conversion and checks) to compute Shanks' 707 places. How, then, did the digit hunters of the 18th and 19th centuries accomplish these feats? For the most part, apparently, by covering acres of paper for months and years; but at least one of them, Strassnitzky, used a human computer, a calculating prodigy.

SEVERAL such phenomenal calculating prodigies have been known, and though their power is little understood, it is clear that they share with the computer its two basic abilities: the rapid execution of arithmetical operations, and the storage (memory) of vast amounts of information. Some of them also have an additional gift: They can recognize large numbers of objects without counting them. This can be done by you or me if that number is three or four; and if the objects are arranged in certain patterns, we can recognize six or ten, perhaps more. But Johann Dase, the man who calculated the 200 decimal places of π in less than two months, could give the number of sheep in a flock, books in a case, etc., after a single glance (up to 30); after a second's look at some dominos, he gave the sum of their points correctly as 117; and when shown a randomly selected line of print, he gave the number of letters correctly as 63.

Truman Henry Safford (1836-1901) of Royalton, Vermont, could instantly extract the cube root of seven-digit numbers at the age of 10. At the same age, he was examined by the Reverend H.W. Adams, who asked him to square, in his head, the number

365,365,365,365,365,365.

Thereupon, reports Dr. Adams,

> He flew around the room like a top, pulled his pantaloons over the tops of his boots, bit his hands, rolled his eyes in their sockets, sometimes smiling and talking, and then seeming to be in agony, until, in not more than a minute, said he,
>
> 133,491,850,208,566,925,016,658,299,941,583,255! [58]

Truman Safford never exhibited his powers in public. He graduated from Harvard, became an astronomer, and gradually lost the amazing powers he had shown in his youth. None of this is typical for most of the calculating prodigies. Many of them, including our Johann Dase, were *idiot savants*: They were brilliant in rapid computations, but quite dull-witted in everything else, including mathematics. Few of them could even intelligently explain how they performed the calculations, and those that could, revealed clumsy methods.

The Englishman Jedediah Buxton (1707-1772), for example, never learned to read or write, but he could, in his head, calculate to what number of pounds, shillings and pence a farthing (¼ penny) would amount if doubled 140 times. (The number of pounds has 39 digits, and the pound had 20 shillings of 12 pence each.) In 1754 he visited London, where several members of the Royal Society satisfied themselves as to the genuineness of his performances. He was also taken to Drury Lane Theater to see a play, but entirely unaffected by the scene, he informed his hosts of the exact number of words uttered by the various actors, and the number of steps taken by others in their dances. Even more surprisingly, his methods of arithmetic, in the rare cases when he could explain them, were quite clumsy. He never learned, for example, to add powers of ten when multiplying; he called 10^{18} "a tribe," and 10^{36} "a cramp."[59]

Johann Dase, to whom we shall return in a moment, was also an *idiot savant*, and the dull wits of these calculating prodigies are a third property that they share with an electronic computer. The sinister "electronic brains" of science fiction and the boob tube to the contrary, the electronic computer is a moron whose total imbecility can often be quite exasperating. If you forget a line in your program telling it to print the results, it will perform all the complicated computations and then erase them again, handing you a blank sheet of paper for a print-out. If you punch the number three as "3" when it should have been "3.", it will refuse to work the program and instead it will print some gobbledegook like this:

ERROR IN LINE 123. ILLEGAL MIXING OF MODES.
EXECUTION DELETED. TIME 23 SECS.

If the computer is so smart, why does it not put in the one dot instead of churning out all this gibberish? Ask it; but it will just sit there, a moronic heap of wire, semiconductors and tape, and say nothing. For a computer is extremely fast, and it has a vast and rapidly accessible memory, but contrary to popular belief, it is totally without intelligence, slavishly following the rules that were built into it. The most intelligent thing it is capable of doing without the help of its programmers is to go on strike when required to work without air conditioning.

To return to Johann Martin Zacharias Dase. He was born in 1840 in Hamburg, had a fair education and was afforded every opportunity to develop his powers, but made little progress; all who knew him agree that except for calculating and numbers, he was quite dull. He always remained completely ignorant of geometry, and never learned any other language than German. His extraordinary calculating powers were timed by renowned mathematicians: He multiplied two 8-digit numbers in his head in 54 seconds; two 20-digit numbers in 6 minutes; two 40-digit numbers in 40 minutes; and two 100-digit numbers (also in his head!) in 8 hours and 45 minutes. To achieve feats like these, he must have had a photographic memory. His ability to recognize the number of objects without counting them has already been remarked on, and one may speculate that perhaps he achieved

CARL FRIEDRICH GAUSS
(1777–1855)

this also by his fantastic memory, that is, by taking a glance at a flock of sheep, and then rapidly counting them from the photographic image in his mind. He could, for example, in half a second memorize a twelve-digit number and then instantly name the digit occupying a particular position, so that whatever the mechanism that enabled his brain to do this, it must have been close to photography. A fantastic memory is, of course, one of the assets that all of the calculating prodigies had in common. Another calculating prodigy, George Parker Bidder (1806-1878), a British civil engineer, could instantly give a 43-digit number after it had been read to him backwards. He did this, at the age of ten, at a performance; and an hour later, asked whether he had remembered it, he immediately repeated it:

2,563,721,987,653,461,598,746,231,905,607,541,128,975,231.

Johann Dase gave exhibitions of his extraordinary calculating powers in Germany, Austria and England, and it was during an exhibition in Vienna in 1840 that he made the acquaintance with Schulz von Strassnitzky, who urged him to make use of his powers for the calculation of mathematical tables. He was then 16 years old, and gladly agreed to do so, thus becoming acquainted with many famous

PROBLEMATICAL RECREATIONS 518

Father watched Junior in his playpen laying out blocks in a neat row. Something about the sequence of letters astounded him. Was Junior a natural mathematician or was it only coincidence? When Junior added the ninth block, Father gasped. The sequence was mathematically perfect. What was the ninth block, and what was the sequence? —*Contributed by P. J. Foley*

Creative motion control solutions are a specialty of our Louis Allis division, makers of motors, adjustable speed drives, digital control and power conversion equipment for 70 years. They offer precise control through full loop capability. For quick, effective solutions to any problem they use the three elements of motion control: sense and control functions performed by Dynapar digital equipment; regulated power in adjustable speed drives from ⅛ to more than 2500 HP; and motion through a full range of motors, gearmotor products and mechanical, adjustable speed drives. Send for the brochure, "Key Ideas Are In Motion At Louis Allis," 427 E. Stewart St., Milwaukee, Wis. 53201.

ANSWER TO LAST WEEK'S PROBLEM: N cuts can divide a cheese into a maximum of 1/6 (N+1) (N²—N+6) pieces. If N=16, this = 697, which represents FIG.

LITTON INDUSTRIES © Copyright 1970

The eternal fascination of π. A 1970 advertisement[60] coding the digits of π ($A = 1$, $B = 2$, $C = 3$, etc.)

mathematcians of his age, including Carl Friedrich Gauss (1777-1855). When he was 20, Strassnitzky taught him the use of the formula

$$\pi/4 = \arctan(1/2) + \arctan(1/5) + \arctan(1/8),$$

with a series expansion for each arctangent, and this is what he used to calculate π to 205 decimal places (of which all but the last five turned out to be correct). This stupendous task he finished in just under two months.

Then he came back for more. He calculated the natural logarithms of the first 1,005,000 numbers, each to 7 decimal places, which he did in his spare time in 1844-1847 when employed by the Prussian Survey. In the next two years he compiled a table of hyperbolic functions, again in his spare time. He also offered to make tables of the factors of all numbers from 7,000,000 to 10,000,000; and on the recommendation of Gauss, the Hamburg Academy of Sciences agreed to assist him financially so that he could devote himself to this work, but he died in 1861, after he had finished about half of it.

It would thus appear that Carl Friedrich Gauss, who holds so many firsts in all branches of mathematics, was also the first to introduce payment for computer time.

It is an interesting phenomenon that all the digit hunters concentrated on the number π; none ever attempted to find hundreds of decimal places for $\sqrt{2}$ or $\sin 1°$ or $\log 2$.[92] There seems to be something magical about the number π that fascinates people: I have known several people who memorized π (in their adolescence) to 12 and even 25 decimal digits; none of them memorized, say, $\sqrt{2}$. There is no mathematical justification for this, for to calculate or memorize π to many decimal places is the same waste of time as doing this for the square root of two. The reason must be psychological; perhaps the explanation is that $\sqrt{2}$ is not so very different from $\sqrt{3}$, and $\sin 1°$ is not so very different from $\sin 2°$; but π is unique. Or at least people think so at the early age when they are first introduced to this number; if they pursue the matter in higher education, they find that the number of transcendentals is not merely infinite, but indenumerable (see Chapter 16).

Various mnemonic devices are available for cluttering the storage cells of one's brain with the decimal digits of π. In the following sentence,[46] for example, the number of letters in each word represents the successive digits of π:

How I want a drink, alcoholic of course, after the heavy lectures involving quantum mechanics! (3.14159265358979)

In French and German, there are even poems for this purpose. In French, there is the following poem:

Que j'aime à faire apprendre un nombre utile aux sages!
Immortel Archimède, artiste ingénieur,
Qui de ton jugement peut priser la valeur?
Pour moi, ton problème eut de pareils avantages.

This poem evidently inspired some German to the following bombastic lines of Teutonic lyrics:

Dir, o Held, o alter Philosoph, du Riesengenie!
Wie viele Tausende bewundern Geister
Himmlisch wie du und göttlich!
Noch reiner in Aeonen
Wird das uns strahlen
Wie im lichten Morgenrot!

Both poems give π to 29 decimal places. There is also the following German verse giving π to 23 decimal places:

Wie? O! Dies π
Macht ernstlich so vielen viele Müh'!
Lernt immerhin, Jünglinge, leichte Verselein,
Wie so zum Beispiel dies dürfte zu merken sein!

The 32nd decimal digit of π is a zero, so that this kind of poetry is mercifully nipped in the bud.

11

The Last Archimedeans

The reasonable man adapts himself to the world; the unreasonable one persists in trying to adapt the world to himself. Therefore all progress depends on the unreasonable.

GEORGE BERNARD SHAW
(1856-1950)

SOME 1,900 years after Archimedes' *Mensuration of the Circle*, people at last began to wonder if there was not a quicker road to π than comparison of the circle with the inscribed and circumscribed polygons. Two Dutch mathematicians found such a road by elementary mathematics, even before the differential and integral calculus had been invented. It is a tribute to the old genius of antiquity that for almost two millenia his method defied improvement; the numerical evaluation became more accurate as the centuries went by, but the method remained without substantial change.

The first man to challenge Archimedes' polygons was the Dutch mathematician and physicist Willebrord Snellius (1580-1626), professor of mathematics at the University of Leyden, who is today best known for his discovery of the laws of reflection and refraction. (If a light ray is incident onto a plane interface between two media, then the angle of incidence equals the angle of reflection, and the sine of

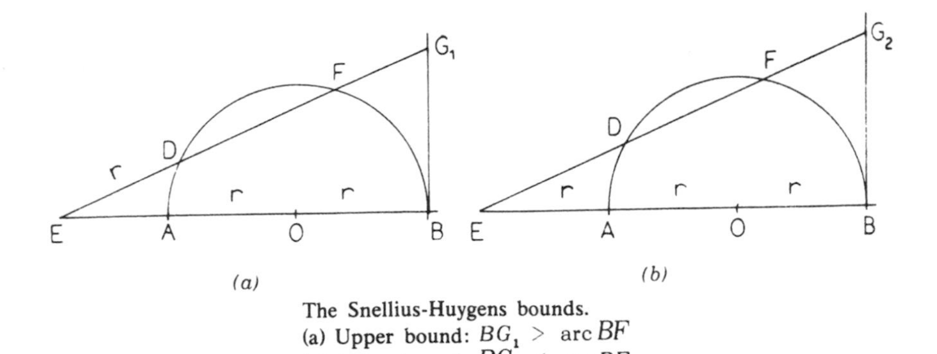

The Snellius-Huygens bounds.
(a) Upper bound: $BG_1 > \text{arc } BF$
(b) Lower bound: $BG_2 < \text{arc } BF$

the angle of incidence to the sine of the angle of refraction equals the refractive index.) In his book *Cyclometricus* (1621), Snellius argues quite correctly that the sides of the inscribed and circumscribed polygons are too widely separated by the arc of the corresponding circle, and that therefore the upper and lower bounds on π resulting from the Archimedean polygons are needlessly far apart. He searched for geometrical configurations that would yield closer bounds to the rectification of an arc, and he found them. They are shown in the figure above. The upper bound (of the discrepancy between circular arc and constructed straight segment) was given by Snellius as follows: Choose a point *D* on the circumference of a given circle (figure *a*) and make *DE* equal to the radius *OA*, where *E* lies on the extension of the diameter *AB*. Let *ED* extended intersect the tangent *BG* at G_1; then

$$BG_1 > \text{arc } BF.$$

The lower bound is given by a construction that we have already discussed in connection with Cardinal Cusanus' approximate rectification of an arc (p. 85). Make *EA* equal to the radius, where *E* lies on *AB* extended; choose any point *F* on the circumference of the circle such that *F* is the second intersection of *EF* and the circle (figure *b*), and let G_2 be the intersection of *EF* and the tangent *BG*; then

$$BG_2 < \text{arc } BF.$$

These bounds were indeed much closer than those of the Archimedean polygons; unfortunately, Snellius was unable to supply a rigorous proof that his assertions were correct. (They were.)

Is it possible to use something that has not been rigorously proven, i.e., that has not been derived by ironclad logic from certain basic assumptions? In physics, the answer is an unqualified yes. We claim that water, all water, boils at 100°C (at standard atmospheric pressure at the elevation of Paris), although we have boiled almost no water whatever compared with all the water in the universe. Our claim is based on the *faith* (yes, it is a faith) that nature is consistent. The Law of the Conservation of Energy is based only on our experience: All of our experience supports it, and none of our experience contradicts it. There is no theoretical reason why tomorrow some scientist should not report an experiment that contradicts it; yet our faith in the consistency or uniformity of nature is so deep that we would probably not believe him. Physics is an inductive science; it finds its laws by generalizing many specific experiences.

The rules of the game in mathematics are slightly different: It is a deductive science, which deduces many specific theorems from a few very general assumptions or axioms. The structure of Euclid's cathedral, for example, is entirely deductive.

And yet the history of mathematics abounds with examples where certain methods and theorems were used long before a rigorous proof was supplied that they are actually true. It is probably no exaggeration to say that most of the progress in mathematics is due to physicists and engineers who discovered certain truths by the physical approach; and only after they had built little houses on sand, the mathematicians supplied foundations of concrete. Euclid did this to the geometry that was, for the most part, known before his day; and this tendency in the development of mathematics can be found in history right down to our own days.

Oliver Heaviside (1850-1925), in his investigations of electrical curcuits, observed certain regularities in the solution of differential equations, and found a method by which difficult differential equations could be converted to simple algebraic equations. "The proof is performed in the laboratory," he proclaimed in the face of the wincing mathematicians. For a quarter of a century, electrical engineers were using Heaviside's "operational calculus," until in the early 1920's the mathematicians discovered that Heaviside's magic could be anchored in the solid ground of an integral transform discovered by Laplace a century before Heaviside. But Heaviside had also used a thing called the delta function, and this, the mathematicians proclaimed, was a monstrosity, for it could be demonstrated that no such function, with the properties Heaviside ascribed to it, could exist. What especially outraged the mathematicians was not so much that electrical engineers continued to use it, but that it would amost always supply the correct result. Not until 1950 did the French mathematician Laurent

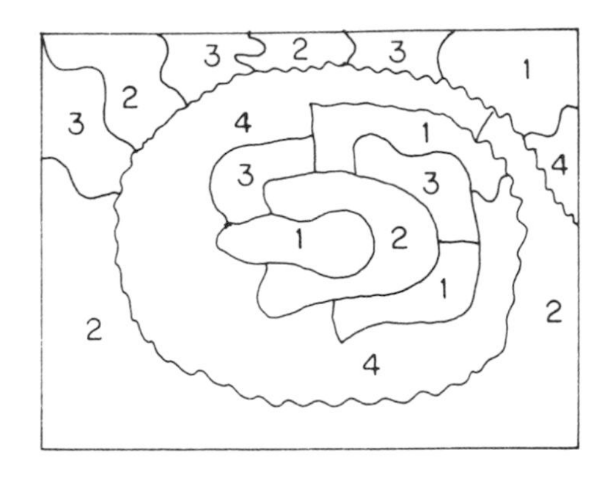

The unproven four-color conjecture. 1, 2, 3, 4 are different colors. No two colors meet at any border between two regions.

Schwartz (1915-) show what was going on; if the delta function could not exist as a function, it could exist as a distribution, and the engineers had been doing things right, after all. Heaviside was one of the unreasonable men in the motto of this chapter.

Even today there are many statements that are "true" by physical standards of experience, but that remain mathematically unproven. Two of the most famous are the Goldbach conjecture and the four-color problem. In 1742, Christian Goldbach (1690-1764), in a letter to Euler, suggested that every even number is the sum of two primes (e.g., $8 = 5 + 3$; $24 = 19 + 5$; $64 = 23 + 41$; etc.). By the physical standards of experience, Goldbach was right, for no one has yet found an even number for which the conjecture does not work out. But that does not constitute a proof, and a mathematical proof has yet to be found. The other well known problem, the so-called four-color problem, is to prove that no matter how a plane is subdivided into non-overlapping regions, it is always possible to paint the regions with no more than four colors in such a way that no two adjacent regions have the same color. By experience, the truth of the assertion is known to every printer of maps (common corners, like Colorado and Arizona, do not count as adjacent.) And no matter how one tries to dream up intertwining states on a fictitious map (see figure above), four colors always appear to be enough. But that is no proof.

Snellius, then, had found two jaws of a vise closer to the circumlar arc than were the Archimedean polygons, and the jaws would close faster on the value of π with every turn, that is, with every doubling of sides of the polygons. For example, for a hexagon, Archimedes obtained the limits $3 < \pi < 3.464$; using the two theorems above, Snellius was able to find the limits $3.14022 < \pi < 3.14160$ for a hexagon, which are closer to π than those that Archimedes had obtained even with the help of a 96-sided polygon. On the contrary, using a 96-sided polygon, Snellius found the limits

$$3.1415926272 < \pi < 3.1415928320.$$

Finally, Snellius verified the decimal places found by Ludolph van Ceulen, and with much less effort than Ludolph had invested.

There is all the difference in the world between Ludolph's digit hunting and Snellius' numerical test. Snellius had found a new method and checked its quality by calculating the decimal digits of π; Ludolph's evaluation to 35 decimal digits by a method known for 1900 years was no more than a stunt.

Snellius' two theorems were rigorously proved by another Dutch mathematician and physicist, Christiaan Huygens (1629-1695). In his remarkable work *De circuli magnitude inventa* (1654), Huygens proved these and 14 more theorems by Euclidean geometry and with Euclidean rigor. Let us denote the area and perimeter of the inscribed polygon by a and p, respectively, and the same for the circumscribed polygon by A and P; let the subscripts 1 and 2 stand for the original polygon and the polygon with double the number of sides, and let a and p (without subscripts) denote the area and perimeter (circumference) of the circle. Then the most important of the 16 theorems proved by Huygens can be written in algebraic notation as follows:

$$a > a_2 + \tfrac{1}{3}(a_2 - a_1) \qquad \text{(V)}$$

$$a < \tfrac{2}{3}A_1 + \tfrac{1}{3}a_1 \qquad \text{(VI)}$$

$$p > p_2 + \tfrac{1}{3}(p_2 - p_1) \qquad \text{(VII)}$$

$$p < \tfrac{2}{3}p_1 + \tfrac{1}{3}P_1 \qquad \text{(IX)}$$

$$p_2 = \sqrt{(P_2 \cdot p_1)} \qquad \text{(XIII)}$$

$$p^3 < p_1^2 \cdot P_1 \qquad \text{(XIV)}$$

$$p > p_2 + (4p_2 + p_1)/(2p_2 + 3p_1) \qquad \text{(XVI, corollary)}$$

The Roman numerals denote the number of Huygens' theorems in the original work. The proofs of these theorems are long and will not be repeated here. (The interested reader can find the original work in *Oeuvres completes de Christiaan Huygens*, Paris, 1888 *et sequ.*, or in

CHRISTIAAN HUYGENS
(1629–1695)

German translation, in Rudio's monograph quoted in the bibliography at the end of this book.) By means of these theorems, Huygens was able to set the bounds

$$3.14159\,26533 < \pi < 3.14159\,26538.$$

Again, this was no mere digit hunting; as Huygens points out in his work, Archimedean polygons of almost 400,000 sides would be needed to obtain π to the same accuracy.

Huygens wrote *De circuli magnitude inventa* as a young man of 25 who had only recently taken up the serious study of mathematics and physics (he had been trained as a lawyer). The work soon lost interest owing to the discovery of better methods supplied by the differential calculus (to which Huygens contributed not only in preparing the ground for it mathematically, but also in other ways: In 1672 he gave Leibniz lessons in mathematics, and in 1674 he transmitted Leibniz's first paper on the differential calculus to the French Academy of Sciences). Huygens was already a contemporary of Newton, and Newton's supreme genius overshadowed everybody else. In a way, Huygens' book on the circle is reminiscent of some 18th century composers such as Rejcha or Vranický; they wrote delighful music, but they were overshadowed by Wolfgang Amadeus Mozart.

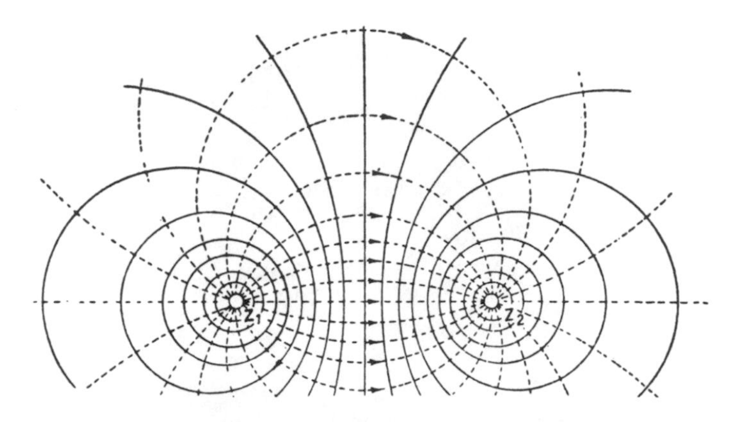

Apollonius circles (full lines). Each circle is the locus of points satisfying the relation $1/a + 1/b = \text{const}$, where a and b are the distances from two fixed points. The broken circles intersect Apollonius' circles at right angles (lines of force). Discovered in the 3rd century B.C.

Even so, Huygens' book documents that European mathematics had at last crawled out of the morass of Roman Empire and Roman Church. Boethius' *Geometry*, written in the last year of the Roman Empire, was a mathematical cookbook; and Pope Sylvester II received requests to explain the strange phenomenon that on doubling the diameter of a sphere, its volume would increase eightfold; but Huygens' treatise had once more attained the high standards of mathematical rigor exacted by the University of Alexandria almost 2,000 years earlier.

Besides, Huygens' made his permanent mark on history as a physicist rather than a mathematician. His many discoveries included a principle of wave motion, which to this day is called Huygens' principle. He could not have known that the antennas of the radio stations tracking man's first flight to the moon would be calculated on the basis of that principle; any more than Apollonius of Perga, in the 3rd century B.C., could have known that the family of circles he discovered (see figure above) would one day be the equipotentials of two parallel, cylindrical, electrical conductors. These are but two of hundreds of examples that one might quote in answer to the question what good comes from exploring the moon or studying non-Euclidean geometry. The question is often asked by people who count cents instead of dollars and dollars instead of satisfaction. Of late, this question is also being asked by the intellectual cripples who drivel

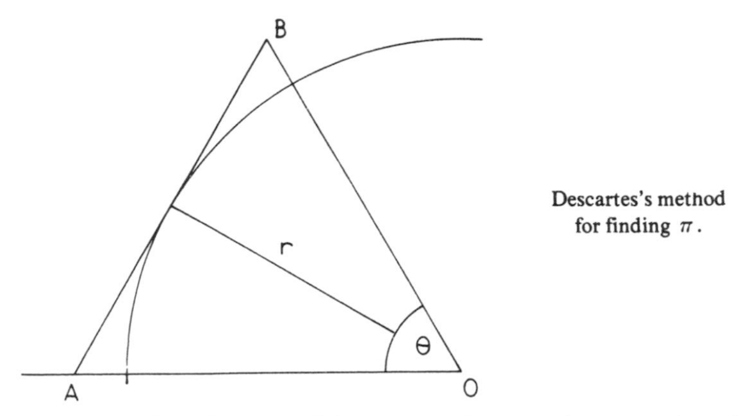

Descartes's method for finding π.

about "too much technology," because technology has wounded them with the ultimate insult: They can't understand it any more.

Some Victorian lady asked Michael Faraday (1791-1867) this question about his discovery of electromagnetic induction, and he answered: "Madam, what is the use of a newborn baby?"

HUYGENS was born three years after Snellius died; a contemporary of both was the famous founder of analytical geometry, René Descartes (1596-1650). He latinized his name to Renatus Cartesius, which is the reason why analytical geometry is sometimes called Cartesian geometry. Descartes was not only a mathematician, but also a physicist and philosopher. He fully accepted the Copernican system, but frightened by the condemnation of Galileo in 1633, he published many of his scientific works in obscure and ambiguous form, and others were not published until after his death.

Among the papers found after his death was some work on the determination of π. His approach is in some ways reminiscent of Viète's, but instead of considering the area of a polygon, he kept its perimeter constant and doubled the sides until it approached a circle.

If the length of the perimeter is L, then as evident from the figure above, in which AB is the side of the polygon,

$$L = 2nr\tan(\theta/2)$$

Hence the radius of the inscribed circle is, after k doublings,

$$r(k) = \frac{L\cot(\theta/2^k)}{2^k n} \tag{1}$$

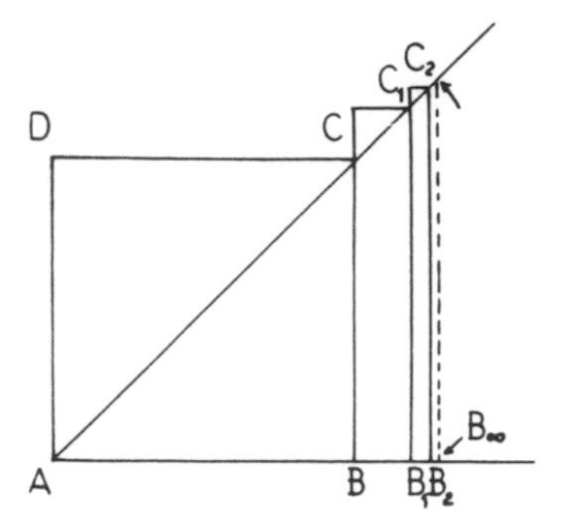

Descartes's construction.

Descartes used a square as a starting point, so that $n = 4$, $\theta = \pi/2$; hence

$$\pi = \lim_{k \to \infty} 2^k \tan(\pi/2^k) \qquad (2)$$

By using the half-angle formula for the tangent, Descartes could have obtained an infinite sequence of operations; however, he did not get as far as (2), but used the equivalent construction shown in the figure above.

Let *ABCD* be a square of given perimeter L, so that $AB = L/4$; extend the diagonal *AC* and determine C_1 so that the area of the rectangle BCC_1B_1 equals ¼ of the area of the square *ABCD* (this can be done by a construction involving only compasses and straightedge). Continue to points C_2, C_3, ..., so that each new rectangle has ¼ the area of the preceding one. Then AB_∞ is the diameter of a circle with circumference L.

Indeed, if $AB_k = x_k$, then $AB = x_0$; then the construction makes

$$x_k(x_k - x_{k-1}) = x_0^2/4^k$$

which is satisfied by

$$x_k = \frac{4x_0}{2^k} \cot \frac{\pi}{2^k},$$

and since $4x_0 = L$, this is equivalent to (1) with $n = 4$.

Huygens' and Snellius' inequality for the lower bound shown in the figure on p. 111 can be used as an approximate construction for rectifying the circle, as we have seen on p. 85-86. Another construction for the approximate rectification of the circle was found by A.A. Kochansky in 1685 (see figure). The angle *BOC* is 30°, *BC* is parallel

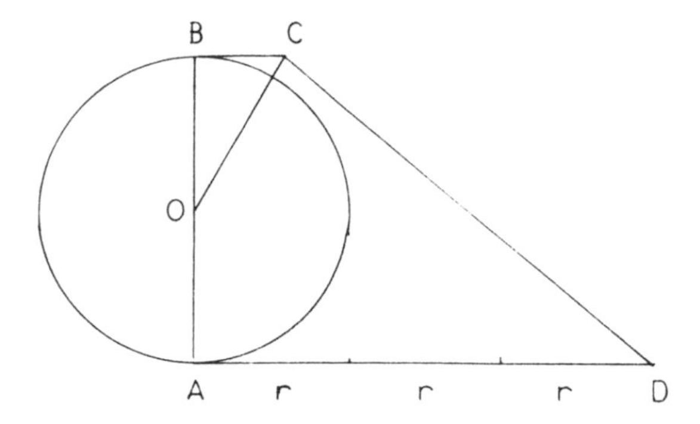

Kochansky's approximate rectification of the circle (1685).

to AD, and the rest is evident from the figure. The length of the segment CD is approximately equal to half the circumference of the circle.

We have

$$BC = r \tan 30^\circ = r/\sqrt{3}$$

and

$$\begin{aligned} CD^2 &= AB^2 + (AD - BC)^2 \\ &= 4r^2 + (3r - r/\sqrt{3})^2 \\ &= (40/3 - 6\sqrt{3})r^2, \end{aligned}$$

so that

$$CD/r = 3.141533\ldots,$$

which differs from π by less than 6×10^{-5}.

A construction corresponding to π correct to 6 decimal places was given by Jakob de Gelder in 1849. It is based on one of the convergents in the expansion of π in a continued fraction, or simply on the fact that $\pi = 355/113$ is correct to six decimal places.

Since $355/113 = 3 + 4^2/(7^2 + 8^2)$, the latter fraction can easily be constructed geometrically (see Gelder's construction overleaf). Let $CD = 1$, $CE = 7/8$, $AF = 1/2$, and let FG be parallel to CD, and FH to EG; then $AH = 4^2/(7^2 + 8^2)$, and the excess of π above 3 is approximated with an error less than one millionth.

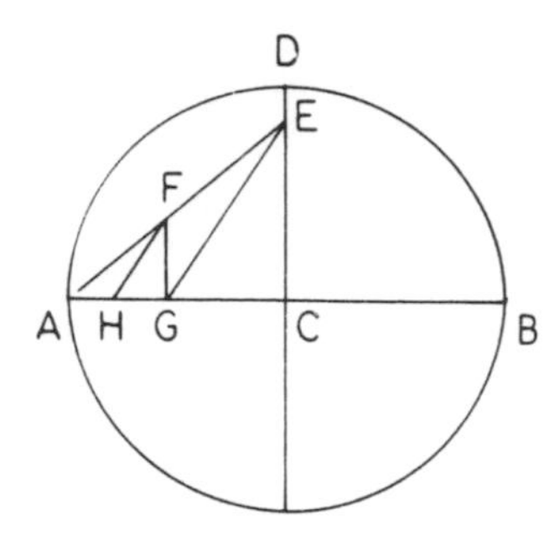

Gelder's construction

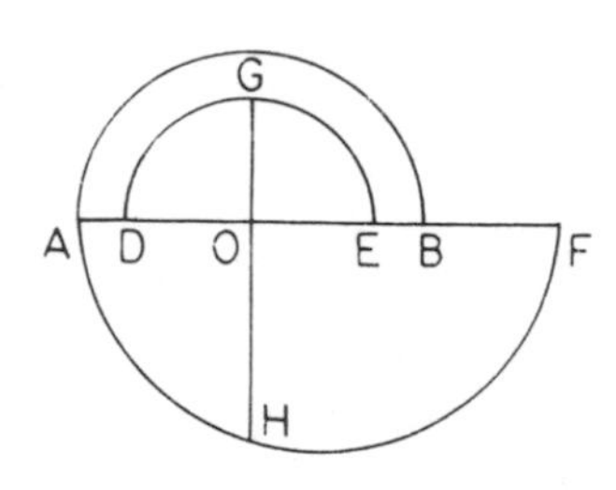

Hobson's construction

A simple construction will also square the circle with an error of less than 2×10^{-5} (see Hobson's construction above). Let $OA = 1$, $OD = 3/5$, $OF = 3/2$, $OE = 1/2$. Describe the semicircles DGE, AHF with DE and AF as diameters, and let the perpendicular to AB through O intercept them in G and H, respectively. Then $GH = 1.77246\ldots$, which differs from $\sqrt{\pi} = 1.77245\ldots$ by less than 2×10^{-5}. Hence GH^2 is very close to the area of the circle with diameter AB. The construction was given by Hobson in 1913.[61]

Other approximate constructions have been collected by M. Simon.[61]

12

Prelude to Breakthrough

I have made this letter longer than usual
because I lacked the time to make it short.
BLAISE PASCAL
(1623-1662)

THE author of the above quotation was one of the most brilliant mathematicians and physicists of the 17th century, or at least, he could have been, had he not flown from flower to flower like a butterfly, finally forsaking the world of mathematics for the world of mysticism. The squaring of the circle or the calculation of π was not one of the flowers Pascal visited, yet his brainstorms were important in preparing the ground for the calculus and the discovery of a new approach to the calculation of π. In his 1658 *Traité des sinus du quart de cercle* (Treatise of the sines of the quadrant of the circle) he came so close to discovering the calculus that Leibniz later wrote that on reading Pascal's work a light suddenly broke upon him.

Pascal began the study of mathematics at age 12. At age 13 he had discovered the pyramid of numbers known as the Pascal triangle (see figure on the next page). Before he was 16, he had discovered Pascal's Theorem (the points of intersection of opposite sides of a hexagon inscribed in a conic are collinear), which became one of the fundamental theorems of projective geometry; by the time he was 17, he had

1

1 1

1 2 1

1 3 3 1

1 4 6 4 1

1 5 10 10 5 1

1 6 15 20 15 6 1

1 7 21 35 35 21 7 1

The Pascal triangle. Each number is the sum of the two above it. The kth line and the mth (oblique) column give the number of combinations of k things taken m at a time, or the coefficient of the mth term of $(x + a)^k$.

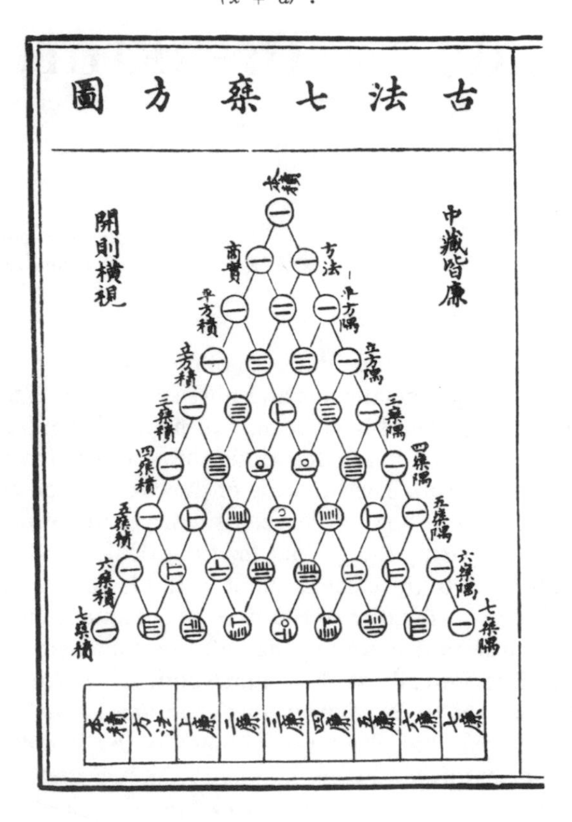

Chinese version of the Pascal triangle, published in 1303, 320 years before Pascal was born.[62]

BLAISE PASCAL
(1623-1662)

used his theorem to derive four hundred propositions in an essay on conic sections. At 19, he invented a calculating machine, some principles of which survived in the desk calculators of a few years ago. In physics, he contributed to several branches, particularly hydrostatics. But on November 23, 1654, Pascal's coach narrowly escaped a fall from a bridge, and experiencing a religious ecstasy, he decided to forsake mathematics and science for theology. Only for a short period thereafter did he devote himself to mathematics: To take his mind off the pains of a toothache, he pondered the problems associated with the cycloid (the curve described by a point on a rolling circle). In this brief period of 1658-59, he achieved more than many others in a lifetime, but soon he returned to problems of theology, and in the last years of his life he mortified his flesh by a belt of spikes round his body, hitting it with his elbow every time a thought entered his mind that was not sufficiently pious.[63]

This, alas, is not untypical for the mental health of infant prodigies in later life. Few are the Mozarts, Alekhines and Gausses, who are born as wonderchildren and retain their extraordinary abilities throughout their lives. Most men of genius matured slowly; neither Newton, nor Euler, nor Einstein, for example, were wonderchildren. They were not even particularly outstanding in mathematics among their schoolmates.

One of the things that Pascal's historic toothache caused was the historic triangle *EEK* in the figure on the next page, which was

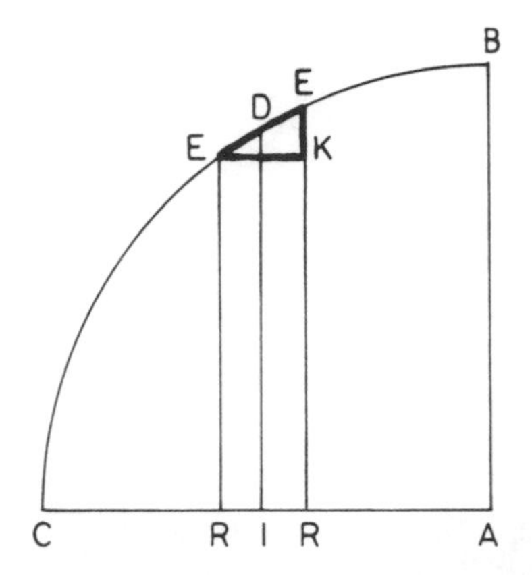

The "historic triangle." Pascal used it to show that

$$DI.EE = RR.AB$$

contained in Pascal's treatise on sines of a quadrant. Pascal used it to integrate (in effect) the functions $\sin^n \phi$, that is, he found the area under the curve of that function. From here it was but a slight step to the integral calculus, and when Leibniz saw this triangle he immediately noticed, as he wrote later, that Pascal's theorems relating to a quadrant of a circle could be applied to any curve. The step that Pascal had missed was to make the triangle infinitely small; but then, all this was a temporary affair started by a toothache. In Boyer's words, "Pascal was without doubt the greatest might-have-been in the history of mathematics."[64]

In applying his triangle to number theory, Pascal also derived a formula, which he expressed verbally, and which is equivalent to

$$\int_0^a x^n\,dx = \frac{a^{n+1}}{n+1} \qquad (n \geqq 0) \tag{5}$$

This formula had also been discovered by several other men of this time, and it was to become very important in the history of π. Pascal was only one of many pioneers who prepared the ground for the calculus. Johann Kepler calculated the area of sectors of an ellipse, which he needed for his Second Law (see figure on opposite page), and he did this by dividing the circle and the ellipse into thin strips, examining the proportions of corresponding areas. Bonaventura Cavalieri (1598-1647) published his *Geometria indivisibilibus continorum* in 1635; here he regarded an area as made up of lines of "indivisibles," and similarly, volumes of indivisible or quasi-atomic areas. By a cumbersome process of comparing the areas to either side

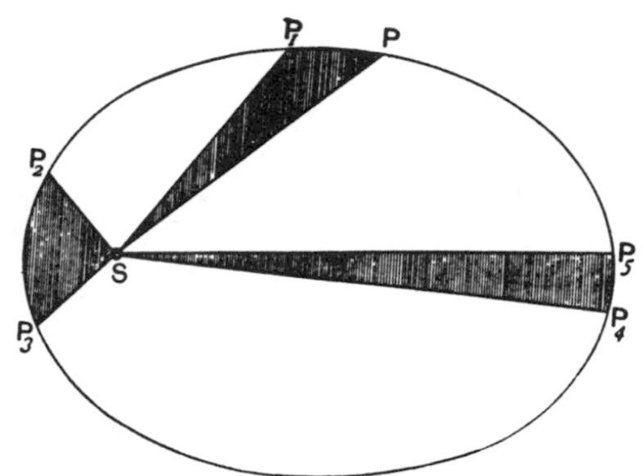

Kepler's second law states that the radius vector (*SP*) of a planet sweeps out equal areas in equal times. Kepler found these areas by analogy with the circle and proportions of corresponding areas.

the diagonal of a parallelogram, he was able to find the equivalent of (1) for $n = 1$, and by an inductive process he generalized this for any n. Evangelista Torricelli (1608-1647), a disciple of Galileo and the inventor of the mercury barometer, came very close to discovering the calculus, as did Gilles Persone de Roberval (1602-1675) and Girard Desargues (1591-1661). Pierre Fermat (1601-1665) not only derived the equivalent of (1), but he could differentiate simple algebraic functions to find their maxima and minima. Isaac Barrow (1630-1677), Newton's professor of geometry at Cambridge, was practically using the differential calculus, marking off "an indefinitely small arc" of a curve and calculating its tangent; but he was a conservative in both mathematics and politics (he sided with the king in the civil war) and worked only with Greek geometry, which made his calculations difficult to follow and his theorems hard to manipulate for further development.

Among these and other pioneers preparing the ground for the calculus, there were at least two who used these little bits and pieces of the coming calculus for deriving the value of π; they were John Wallis (1616-1703) and James Gregory (1638-1675). But before we examine their work, it is instructive to visit Japan, where the value of π may also have been derived by a method of crude integral calculus, resulting in an infinite series which was never used in Europe, presumably because the expressions found by Wallis and Gregory were more advantageous and eliminated the need for the Japanese series.

The evolution of mathematics in Japan was by this time retarded behind that of Europe. In 1722, the Japanese mathematician Takebe was still using a polygon (of 1024 sides) to calculate π to more than

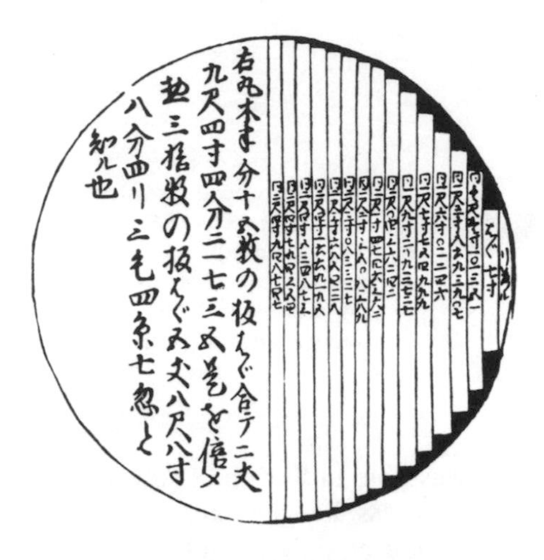

From a book by Sawaguchi Kazayuki (1670),[65] showing early steps in the calculus.

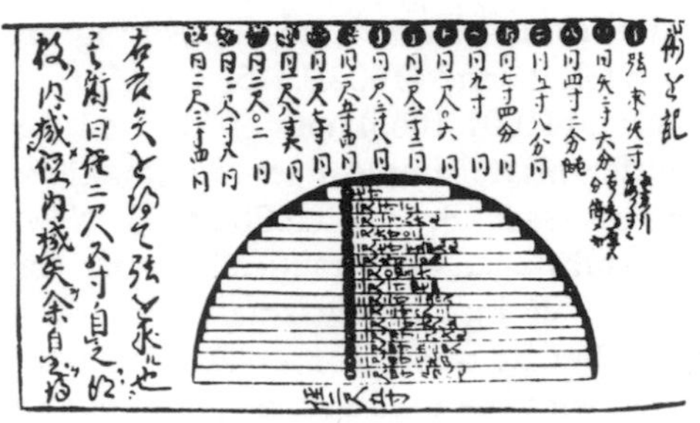

From a book by Machinag and Ohashi (1687).[66]

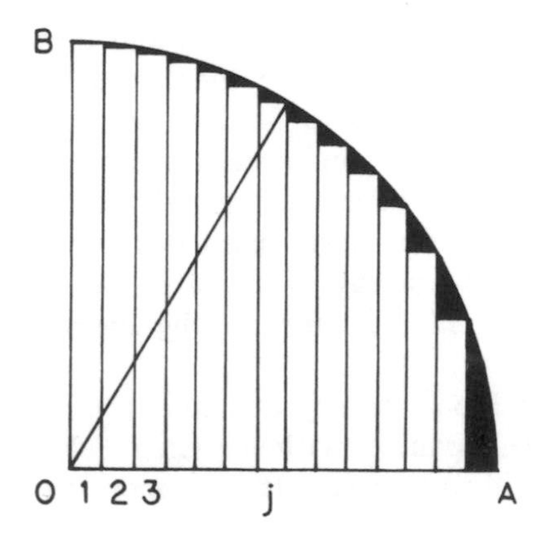

Japanese method of calculating π suggested by figures above.

40 places, although he also found series and continued fractions for π, and also for π^2. At that time Japanese mathematicians rarely gave a clue how they found their results. Perhaps they kept their methods secret, as Tartaglia kept his solution of the cubic equation secret; perhaps they did not have convincing proofs and published "cookbooks" as was the custom in Europe in the Middle Ages. But drawings such as the ones on the opposite page suggest[67] that they may have found a series for π as follows.

Setting the radius OA (see bottom figure on opposite page) equal to unity, the area of the quadrant OAB is $\pi/4$; dividing OA into n intervals, the area of the jth strip is by Pythagoras' Theorem

$$(1/n)\sqrt{1-(j/n)^2} \tag{2}$$

The total area of the strips will tend to the area of the quadrant for $n \to \infty$, yielding

$$\pi = \lim_{n\to\infty} \frac{4}{n^2}\sum_{j=0}^{n} \sqrt{n^2 - j^2} \tag{3}$$

Without converting this expression to an integral, the Japanese could thus obtain π to any desired degree of accuracy by choosing n sufficiently large, at least in theory; in practice, the series would converge very slowly and, like Viète's formula, it would involve the inconvenient extraction of square roots.

Whether or not the Japanese actually used this method is not at all certain; but certainly John Wallis used a very similar approach. Wallis was also looking for the area of a quadrant of a circle. He did not convert (3) to an integral, but he did what amounted to the same thing; he looked for the area under the circular arc AB whose equation was known from Descartes' coordinate geometry. In modern symbols, therefore, he started from the equation

$$\int_0^1 \sqrt{1-x^2}\, dx = \pi/4 \tag{4}$$

However, he did not yet have a method to evaluate the integral on the left, for neither Newton nor Leibniz had discovered the rules of the integral calculus yet. He was not even able to expand the integrand by the binomial theorem and integrate [using the Cavalieri-Fermat-

Pascal formula (1)] term by term, for the binomial theorem at that time was known only for integral powers. How he did it is a long and painful story involving a cumbersome series of interpolations and inductive procedures; but he was able, in his *Arithmetica infinitorum* (1655), to derive the famous formula which bears his name, and which can be written as

$$\pi = 2 \frac{2.2.4.4.6.6...}{1.3.3.5.5.7...} \tag{5}$$

An English translation of this part of his book is readily available[68] and the reader can look up the way in which Wallis sweated it out. However, today we can shortcut his procedure by two simple integrals,

$$\int_0^{\pi/2} \sin^{2m} x \, dx = \frac{1.3.5...(2m-1)}{2.4.6...2m} \frac{\pi}{2} \tag{6}$$

$$\int_0^{\pi/2} \sin^{2m+1} x \, dx = \frac{2.4.6...2m}{1.3.5...(2m+1)} \tag{7}$$

which are both easily derived by iterated integration by parts. Since the limit of the ratio of the integrals (6) and (7) for $m \to \infty$ is one, Wallis' formula follows immediately.

The Wallis formula is a great milestone in the history of π. Like Viète, Wallis had found π in the form of an infinite product, but he was the first in history whose infinite sequence involved only rational operations; there were no square roots to obstruct the numerical calculation as was the case for Viète's formula and in the Archimedean method.

The man who achieved this historic result was Savillian Professor of Geometry at Oxford, a highly and broadly educated man who had graduated from Cambridge in medicine and philosophy and who, apart from his many important contributions to mathematics, published a grammar of English (*Grammatica Linguae Anglicanae*, 1652) and translated many important works from Greek, including Archimedes' *Measurement of the Circle* (1676). Apart from deriving an infinite product for π, he had at least one other thing in common with Viète, and that was deciphering the secret codes of enemy messages, in Wallis' case for the Parliamentarians in the Civil War. In appreciation of this, Cromwell appointed him Professor at Oxford, in spite of his Royalist leanings. This may sound a little confused, but it must be remembered that the English Revolution developed into a three-cornered fight between royalists, parliamentarians and the army under Oliver Cromwell. All modern revolutions pitted the people against a tyrannical regime, and most ended up by substituting one tyrant for another. The English rose against Charles I, but found themselves under Cromwell; the French rose against the *ancien régime*, and found themselves under Robespierre and Napoleon; and the Russians rose against the Tsar to find themselves ruled by the Soviet Jenghis Khans. Wallis found himself ruled by Charles II after the Restoration in 1660, but he was reappointed as professor at Oxford in the same year, and his many activities included the duty of chaplain to that unusual monarch who loved women more than he loved power.

Of the many mathematical works published by John Wallis, the *Arithmetica infinitorum* is the most famous. The Greeks, perhaps because of Zeno's paradoxes, had a horror of the infinite, and with the exception of Archimedes, who toyed with it in *The Method*, they preferred to leave it alone. Cavalieri and the others mentioned above began to attack the infinite, but Wallis was the man who found the right doors, even if he could not yet open them. His *Arithmetica infinitorum* was called "a scab of symbols" by the well known philosopher Thomas Hobbes (1588-1679), who claimed to have squared the circle, and whose philosophy advocated that human beings surrender their individual rights to constitute a state under an absolute sovereignty. Wallis could well afford to ignore Hobbes, and we shall do the same.

WILLIAM, Viscount Brouncker (ca.1620-1684), the first president of the Royal Society, manipulated Wallis' result into the form of a continued fraction.

Continued fractions are part of the "lost mathematics," the mathematics now considered too advanced for high school and too elementary for college. Continued fractions are useful, for example, for solving Diophantine equations. But in recent times a simple way has been found to ensure that continued fractions are not needed for the solution of Diophantine equations: The latter have been kicked out of the high school curriculum also.

Consider the equation

$$x^2 - x - 1 = 0 \tag{8}$$

and write it in the form

$$x = 1 + 1/x \tag{9}$$

Substituting this expression for x on the right side of the same expression, we have

$$x = 1 + \frac{1}{1 + 1/x} \tag{10}$$

If now (9) is substituted for x in the right side of (10), and substituted again every time x appears, we obtain the continued fraction

$$x = 1 + \cfrac{1}{1 + \cfrac{1}{1 + \cfrac{1}{1 + \ldots\text{etc.}}}} \tag{11}$$

As can be checked from the original equation (8), the limit of (11) is the irrational number

$$x = \tfrac{1}{2}(1 + \sqrt{5}) = 1.61803\ldots, \tag{12}$$

and this number can now be approximated by a rational fraction as closely as desired on cutting off the continued fraction (11) at a correspondingly advanced point. The rational fractions obtained by cutting off the process at successive steps are called convergents; for example, the convergents of (11) are

$$1,\ 2,\ 3/2,\ 5/3,\ 8/5,\ 13/8,\ 21/13,\ 34/21,\ 55/34,\ 89/55,\ 144/89,\ \ldots\ \text{etc.} \tag{13}$$

The process converges fairly quickly; for example, the last fraction in (13) is

$$144/89 = 1.61798,$$

which agrees with (12) to four significant figures. Also, there is a quicker way of obtaining the convergents (13) than to worry them out from (11). This is explained in any textbook on continued fractions.[69]

The result that Brouncker transformed into a continued fraction was the one actually given by Wallis, which was $4/\pi = \ldots$ rather than $\pi = \ldots$ as given here by (5). The continued fraction that Brouncker obtained was the pretty expression

$$4/\pi = 1 + \cfrac{1^2}{2 + \cfrac{3^2}{2 + \cfrac{5^2}{2 + \cfrac{7^2}{2 + \cfrac{9^2}{2 + \ldots}}}}} \tag{14}$$

with convergents 1, 3/2, 15/13, 105/76, 945/789, . . .

How Brouncker obtained this result is anybody's guess; Wallis proved its equivalence with his own result (5), but his proof is so cumbersome that it almost certainly does not reflect Brouncker's derivation. Brouncker's result was later also proved by Euler (1775), whose proof amounted to the following. Consider the convergent series

$$S = a_0 + a_1 + a_1 a_2 + a_1 a_2 a_3 + a_1 a_2 a_3 a_4 + \ldots$$

which is easily seen to be equivalent to the continued fraction

$$S = a_0 + \cfrac{a_1}{1 - \cfrac{a_2}{1 + a_2 - \cfrac{a_3}{1 + a_3 - \ldots}}}$$

Now consider the series

$$\arctan x = x - x^3/3 + x^5/5 - x^7/7 + \ldots$$

and set $a_0 = 0$, $a_1 = x$, $a_2 = -x^2/3$, $a_3 = -3x^2/5, \ldots$; then

$$\arctan x = \cfrac{x}{1 + \cfrac{x^2}{3 - x^2 + \cfrac{9x^2}{5 - 3x^2 + \cfrac{25x^2}{7 - 5x^2 + \ldots}}}}$$

and on setting $x = 1$ (which makes $\arctan x = \pi/4$), Brouncker's result (14) follows immediately.

NEWTON later based his work on that of Wallis and Barrow, and it seems that he was not well aware of the work done by the young Scotsman James Gregory (1638-1675), a leading contributor to the discovery of the differential and integral calculus. Gregory was a mathematician, occasionally dabbling in astronomy, who had studied mathematics at Aberdeen, and later in Italy (1664-68). He worked on problems far ahead of his time; for example, in Italy he wrote the *Vera circuli et hyperbolae quadratura* (True quadrature of the circle and the hyperbola), which included the basic idea of the disctinction between algebraic and transcendental functions, and he even attempted to prove the transcendence of π, a task that was not crowned by success until 1882. He was familiar with the series expansion of $\tan x$, $\sec x$, $\arctan x$, $\operatorname{arcsec} x$, and the logarithmic series. In 1668 he returned to Scotland, where he became Professor of Mathematics at St. Andrews University, and in 1674 he was appointed to the first Chair of Mathematics at the University of Edinburgh, but he died suddenly the next year at the age of only 36.

Among the many things that Gregory discovered, the most important for the history of π is the series for the arctangent which still bears his name. He found that the area under the curve $1/(1 + x^2)$ in the interval $(0, x)$ was $\arctan x$; in modern symbols,

$$\int_0^x \frac{dx}{1 + x^2} = \arctan x \tag{15}$$

By the simple process of long division in the integrand and the use of Cavalieri's formula (1), he found the Gregory series

$$\arctan x = x - x^3/3 + x^5/5 - x^7/7 + \ldots \tag{16}$$

From here it was a simple step to substitute $x = 1$; since $\arctan(1) = \pi/4$, this yields

$$\pi = 4(1 - 1/3 + 1/5 - 1/7 + \ldots), \tag{17}$$

which was the first infinite series ever found for π.

Gregory discovered the series (16) in 1671, reporting the discovery in a letter of February 15, 1671, without derivation.[70] Leibniz found the

GOTTFRIED WILHELM LEIBNIZ
(1646-1716)

series (16) and its special case (17) in 1674 and published it in 1682, and the series (17) is sometimes called the Leibniz series. Gregory did not mention the special case (17) in his published works. Yet it is unthinkable that the discoverer of the series (16), a man who had worked on the transcendence of π, should have overlooked the obvious case of substituting $x = 1$ in his series. More likely he did not consider it important because its convergence (a concept also introduced by Gregory) was too slow to be of practical use for numerical calculations. If this was the reason why he did not specifically mention it, he was of course quite right. It was left to Newton to find a series that would converge to π more rapidly.

13

NEWTON

Nature and nature's laws lay hid in night,
God said, Let Newton be, and all was light.
ALEXANDER POPE
(1688-1744)

THERE had never been a scientist like Newton, and there has not been one like him since. Not Einstein, not Archimedes, not Galileo, not Planck, not anybody else measured up to anywhere near his stature. Indeed, it is safe to say that there can never be a scientist like Newton again, for the scientists of future generations will have books and libraries, microfilms and microfiches, magnetic discs and other computerized information to draw on. Newton had nothing, nothing except Galileo's qualitative thoughts and Kepler's laws of planetary motion. With little more than that to go on, Newton formulated three laws that govern all motion in the universe: From the galaxies in the heavens to the electrons whirling round atomic nuclei, from the cat that always falls on its feet to the gyroscopes that watch over the flight of space ships. His laws of motion have withstood the test of time for three centuries. The very concepts of space, time and mass have crumbled under the impact of Einstein's theory of relativity; age-old prejudices of cause, effect and certainty were destroyed by quantum mechanics; but Newton's laws have come through unscathed.

SIR ISAAC NEWTON
From a bust in the Royal Observatory, Greenwich.

Yes, that is so. Contrary to widespread belief, Newton's laws of motion are not contradicted by Einstein's Theory of Special Relativity. Newton never made the statement that force equals mass times acceleration. His Second Law says

$$F = d(mv)/dt$$

and Newton was far too cautious a man to take the *m* out of the bracket. When mass, in Einstein's interpretation, became a function of velocity, not an iota in Newton's laws needed to be changed. It is therefore incorrect to regard relativistic mechanics as refining or even contradicting Newton's laws: Einstein's building is still anchored in the three Newtonian foundation stones, but the building is twisted to accommodate electromagnetic phenomena as well. True, Newton's law

PHILOSOPHIÆ

NATURALIS

PRINCIPIA

MATHEMATICA.

AUCTORE

ISAACO NEWTONO, Eq. Aur.

Editio tertia aucta & emendata.

LONDINI:

Apud Guil. & Joh. Innys, Regiæ Societatis typographos.
MDCCXXVI.

Title page of the third edition of the *Principia* (1726) as reprinted in 1871.

of gravitation turned out to be (very slightly) inaccurate; but this law, even though it led Newton to the discovery of the foundation stones, is not a foundation stone itself.

Newton's achievement in discovering the differential and integral calculus is, in comparison, a smaller achievement; even so, it was epochal. As we have seen, the ground was well prepared for its discovery by a sizable troop of pioneers. Leibniz discovered it independently of Newton some 10 years later, and Newton would not have been the giant he was if he had overlooked it. For Newton overlooked nothing. He found all the big things that were to be found in his time, and a host of lesser things (such as a way to calculate π) as well. How many more his ever-brooding mind discovered, we shall never know, for he had an almost obsessive aversion to publishing his works. The greatest scientific book ever published, his *Principia*, took definite shape in his mind in 1665, when he was 23; but he did not commit his

EDMOND HALLEY
(1656-1742)
Of his many great discoveries, the greatest was the discovery of the *Principia* in Newton's drawer.

theories to paper until 1672-74. Whether he wrote them down for his own satisfaction or for posterity, we do not know, but the manuscript (of Part I) lay in his drawer for ten more years, until his friend Edmond Halley (1656-1742) accidentally learned of its existence in 1684. Halley was one of the world's great astronomers; yet his greatest contribution to science was persuading Newton to publish the *Principia*, urging him to finish the second and third parts, seeing them through the press, and financing their publication. In 1687 this greatest of all scientific works came off the press and heralded the birth of modern science.

Isaac Newton was born on Christmas Day, 1642, in a small farm house at Woolsthorpe near Colsterworth, Lincolnshire. At Grantham, the nearest place that had a school, he did not excel in mathematics in the dazzling way of the wonderchildren Pascal or Gauss, but his schoolmaster, Mr. Stokes, noticed that the boy was bright. If there was any omen of young Isaac's future destiny, it must have been his habit of brooding. Going home from Grantham, it was usual to dismount and lead one's horse up a particularly steep hill. But Isaac would occasionally be so deeply lost in meditation that he would forget to remount his horse and walk home the rest of the way.

When he finished school, there came the great turning point of Newton's career. His widowed mother wanted him to take over the farm, but Stokes was able to persuade her to send Isaac to Cambridge, where he was first introduced to the world of mathematics.

The Manor House at Woolsthorpe. Birthplace of Isaac Newton (1642) and birthplace of the calculus (1665-1666).[71]

What if Stokes had not been able to persuade Mrs. Newton? There are many similar questions. What if Gauss's teacher had not prevailed over Gauss's father who did not want his son to become an "egg-head"? What if G.H. Hardy had paid no attention to the mixture of semi-literate and brilliant mathematical notes sent to him by an uneducated Indian named Ramanujan? The answer, no doubt, is that others would eventually have found the discoveries of these men. Perhaps this thought is some consolation to you, but it leaves me very cold. How many little Newtons have died in Viet Nam? How many Ramanujans starve to death in India before they can read or write? How many Lobachevskis languish in Siberian concentration camps?

However, Newton did go to Cambridge, where he was very quickly through with Euclid, and soon he mastered Descartes' new geometry. By the time he was twenty-one, he had discovered the binomial theorem for fractional powers, and had embarked on his discovery of infinite series and "fluxions" (derivatives). Soon he was correcting, and adding to, the work of his professor and friend, Isaac Barrow. In 1665 the Great Plague broke out, in Cambridge as well as London, and the university was closed down. Newton returned to Woolsthorpe for the rest of the year and part of the next. It is most probable that during this time, when he was twenty-three, with no one about but his mother to disturb his brooding, Newton made the greater part of his

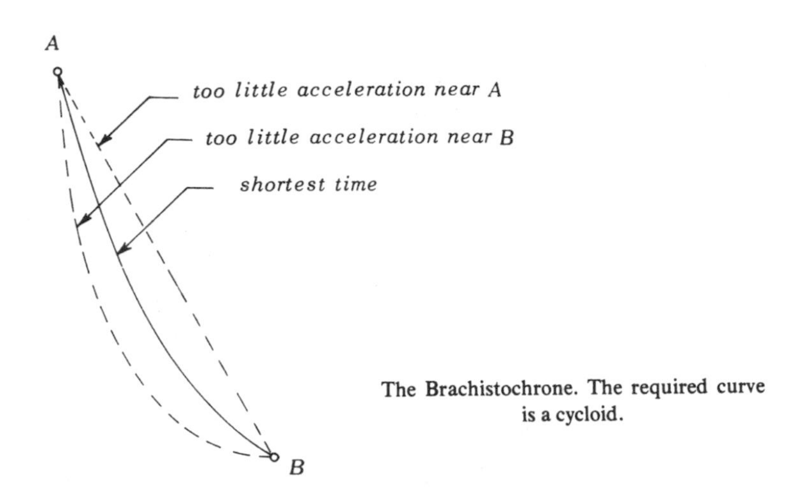

The Brachistochrone. The required curve is a cycloid.

vast discoveries. "All this was in the two plague years 1665 and 1666," he reminisced in old age, "for in those days I was in the prime of my age of invention, and minded mathematics and [natural] philosophy more than at any time since."[72] Asked how he made his discoveries, he answered, "By always thinking unto them," and on another occasion, "I keep the subject constantly before me and wait till the first dawnings open little by little into the full light." Newton retained these great powers of concentration throughout his life. He succeeded Barrow as Lucasian Professor of Mathematics at Cambridge (1669), and relinquished this post to become Warden of the Mint (1696) and later (1699) Master of the Mint; in 1703 he was elected President of the Royal Society, a position which he held until his death in 1726. In his later years he spent much time on non-scientific activity, but remained as astute a mathematician as ever, amazing men by the ease with which he solved problems set up to challenge him.

In 1697, for example, Jean Bernoulli I (1667-1748) posed a problem that was to become famous in the founding of the Calculus of Variations: What is the curve joining two given points (see figure above) such that a heavy particle will move along the curve from the upper to the lower point in minimum time? The problem is so difficult that it is not, for example, usually included in today's undergraduate engineering curriculum. It was received by the Royal Society and handed to Newton in the afternoon; he returned the solution the next morning, and according to John Conduitt (his niece's husband), he solved it before going to bed! The solution was sent to

Jean Bernoulli without signature, but on reading it he instantly recognized the author, as he exclaimed, *tanquam ex ungue leonem* (as the lion is known by its claw).

FOR a giant like Newton, the calculation of π was chickenfeed, and indeed, in his *Method of Fluxions and Infinite Series*, he devotes only a paragraph of four lines to it, apologizing for such a triviality with a *by the way* in parentheses — and then gives its value to 16 decimal places.

Newton wrote this treatise in Latin and it did not appear until after his death in 1742; an English translation appeared earlier in 1737 (printed, the title page tells us, by J. Millan "next to Will's Coffee House at the Entrance to Scotland Yard"). However, he brooded out the major parts of this and other treatises on the differential and integral calculus (including infinite series, which Newton never separated from the calculus) during the Plague Years 1665-66 at Woolsthorpe. His method enabled him to expand a function, its integral or derivative in an infinite series. Where Gregory had found four series for the trigonometric functions, Newton could choose one that would make the calculation of π as rapid as possible.

The Gregory-Leibniz series

$$\pi/4 = 1 - 1/3 + 1/5 - 1/7 + \dots \qquad (1)$$

was theoretically interesting, as it pointed the way to a completely new approach to calculating π. But for numerical calculations it was practically useless, for its convergence was so slow that 300 terms were insufficient to obtain even two decimal places, and two decimal places were less accurate than 3 1/7, the value Archimedes had obtained 2,000 years earlier. De Lagny, whom we have met earlier as the digit hunter who calculated 117 decimal places, found that to obtain 100 decimal places, the number of necessary terms would be no less than 10^{50} !

Newton had found a way to calculate the fluxion (derivative) of a fluent (variable), and conversely, to find the Flowing Quantity from a given fluxion (to integrate). He also showed that this amounted to finding the area under a curve (whose equation was given by the fluxion). He thus found (in modern symbols)

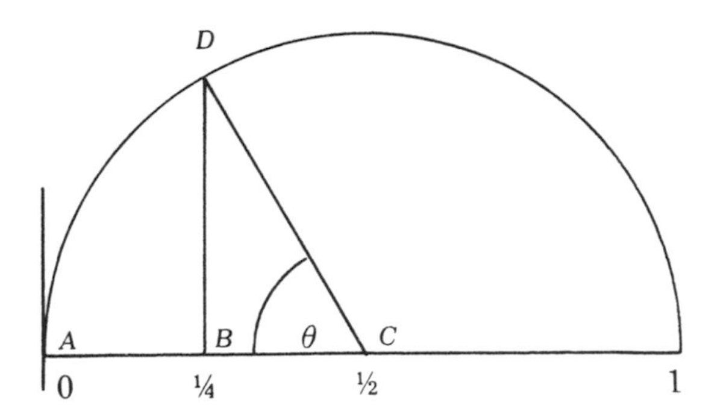

Newton's method of calculating π.

$$\int \frac{dx}{\sqrt{1 - x^2}} = \arcsin x \qquad (2)$$

or using his discovery of the binomial theorem,

$$\int \frac{dx}{\sqrt{1 - x^2}} = \int \left(1 + \frac{1}{2} x^2 + \frac{1.3}{2.4} x^4 + \frac{1.3.5}{2.4.6} x^6 + \dots\right) dx$$

so that integrating term by term and using (2),

$$\arcsin x = x + \frac{1}{2} \frac{x^3}{3} + \frac{1.3}{2.4} \frac{x^5}{5} + \dots \qquad (3)$$

Substituting $x = ½$, which makes $\arcsin(½) = \pi/6$, this yields the series

$$\pi = 6\left(\frac{1}{2} + \frac{1}{2.3.2^3} + \frac{1.3}{2.4.5.2^5} + \dots\right) \qquad (4)$$

which converges incomparably more quickly than the Gregory-Leibniz series.

That is what most history books say on Newton's method of obtaining π. In looking up the original work,[73] however, we find that Newton used a slightly different method. He considered a circle (along with an hyperbola which need not interest us here) whose equation is

$$y = \sqrt{x - x^2} \tag{5}$$

i.e., a circle (see figure on p. 141) with radius ½ centered at $y = 0$, $x = ½$. Then the circular segment ADB has area

$$a = \int_0^{1/4} \sqrt{(x - x^2)}\,dx = \int_0^{1/4} \sqrt{x}\,\sqrt{(1 - x)}\,dx$$

$$= (2/3)\,x^{3/2} - (1/5)\,x^{5/2} - (1/28)\,x^{7/2} - (1/72)x^{9/2}\Big|_0^{1/4}$$

$$= 2/(3.2^3) - 1/(5.2^5) - 1/(28.2^7) - 1/(72.2^9) - \ldots \tag{6}$$

where Newton again used the binomial theorem.

On the other hand, the segment ABD equals the sector ACD less the triangle BCD, and since $CD = 1$, $BD = \sqrt{3/4}$, Newton found

$$a = \pi/24 - \sqrt{3}/32 \tag{7}$$

On comparing (6) and (7), this yields

$$\pi = \frac{3\sqrt{3}}{4} + 24\left(\frac{1}{12} - \frac{1}{5.2^5} - \frac{1}{28.2^7} - \frac{1}{72.2^9} - \ldots\right) \tag{8}$$

and this is how Newton obtained his value of π (see opposite page). Twenty-two terms were sufficient to give him 16 decimal places (the last was incorrect because of the inevitable error in rounding off). A far cry from the Archimedean polygon, where 96 sides (and extracting square roots four times over) yielded only two decimal places!

Actually, Newton was calculating something else, and π appeared only as an incidental fringe benefit in the calculation. But the crumbs dropped by giants are big boulders.

This was one of the calculations he performed during the Plague Years 1665-6 in Woolsthorpe. Later he wrote, "I am ashamed to tell you to how many figures I carried these computations, having no other business at the time."

Scarce any difficulty can occur to any one, who is to undertake ſuch a computation in Numbers, after the value of the Area is obtained in ſpecies. Yet for the more compleat illuſtration of the foregoing doctrine, I ſhall add an Example or Two.

Let the Hyperbola AD be propoſed, whoſe equation is $\sqrt{x+xx}=z$, its vertex being at A, and each of its Axes equal to unity; from what goes before, its Area ADB $=\frac{2}{3}x^{\frac{3}{2}}+\frac{1}{5}x^{\frac{5}{2}}-\frac{1}{28}x^{\frac{7}{2}}+\frac{1}{72}x^{\frac{9}{2}}-\frac{5}{704}x^{\frac{11}{2}}$, &c. that is, $x^{\frac{1}{2}}$ into $\frac{2}{3}x+\frac{1}{5}x^2-\frac{1}{28}x^3+\frac{1}{72}x^4-\frac{5}{704}x^5$, &c. which ſeries may be infinitely produced by multiplying the laſt term continually by the ſucceeding terms of this progreſſion, $\frac{1.3}{2.5}x.\ \frac{-1.5}{4.7}x.\ \frac{-3.7}{6.9}x.\ \frac{-5.9}{8.11}x.\ \frac{-7.11}{10.13}x$, &c. that is, the firſt term $\frac{2}{3}x^{\frac{3}{2}}$ multiplied by $\frac{1.3}{2.5}x$, makes the ſecond term $\frac{1}{5}x^{\frac{5}{2}}$; which multiplied by $\frac{-1.5}{4.7}x$, makes the third term $\frac{-1}{28}x^{\frac{7}{2}}$; which multiplied by $\frac{-3.7}{6.9}$, makes the fourth term $+\frac{1}{72}x^{\frac{9}{2}}$. And ſo on *ad infinitum.*

Now let AB be aſſumed of any length, ſuppoſe $\frac{1}{4}$, and writing this number for x, and its root $\frac{1}{2}$ for $x^{\frac{1}{2}}$, the firſt term $\frac{2}{3}x^{\frac{3}{2}}$ or $\frac{2}{3}\times\frac{1}{8}$ being reduced to a decimal fraction, becomes 0.083333333, &c. this into $\frac{1.3}{2.5.4}$ makes 0 . 00625 the ſecond term; this into $\frac{-1.5}{4.7.4}$ makes 0 . 0002790178, &c. the third term. And ſo on for ever. But the

S

the terms thus reduced by degrees, I diſpoſe into Two Tables; the affirmative terms in One, and the Negative in Another, and add them up as you ſee here.

Affirmative	Negative
+ 0 . 0833333333333333	− 0 . 0002790178571429
62500000000000	34679066051
271267361111	834465027
5135169396	26285354
144628917	961296
4954581	38676
190948	1663
7963	75
352	4
16	0 . 0002825719389575
1	+ 0 . 0896109885646618
0 . 0896109885646618	0 . 0893284166257043

Then from the ſum of the affirmative, I take the ſum of the negative terms, and there remains 0.0893284166257043 for the quantity of the Hyperbolick Area A*d*B which was to be found.

Let the Circle A*d*F [*See the ſame Fig.*] be propoſed, which is expreſſed by the equation $\sqrt{x-xx}=z$, whoſe diameter is unity; and from what goes before its Area A*d*B will be $\frac{2}{3}x^{\frac{3}{2}}-\frac{1}{5}x^{\frac{5}{2}}-\frac{1}{28}x^{\frac{7}{2}}-\frac{1}{72}x^{\frac{9}{2}}$, &c. in which ſeries, ſince the terms do not differ from the terms of the ſeries which above expreſſed the Hyperbolick Area, except in the ſigns + and −; nothing elſe remains to be done, than to connect the ſame numeral terms with their ſigns; that is, by ſubtracting the connected ſums of both the forementioned Tables, 0.0898935605036193, from the firſt term doubled 0.1666666666666666, &c. and the remainder 0.0767731061630473 will be the portion A*d*B of the Circular Area, ſuppoſing AB to be a fourth part

part of the Diameter. And hence we may obſerve, that though the Areas of the Circle and Hyperbola are not expreſſed in a Geometrical conſideration, yet each of them is diſcovered by the ſame Arithmetical computation.

The portion of the Circle A*d*B being found, from thence the whole Area may be derived. For the radius *d*C being drawn, multiply B*d* or $\frac{1}{4}\sqrt{3}$ into BC or $\frac{1}{4}$, and one half of the product $\frac{1}{32}\sqrt{3}$, or 0 . 0541265877365275 will be the value of the Triangle C*d*B; which added to the Area A*d*B, will give the Sector AC*d*, 0 . 1308996938995747, the Sextuple of which 0 . 7853981633974482 is the whole Area.

And hence (by the way) the length of the Circumference will be 3 . 1415926535897928, which is found by dividing the Area by a fourth part of the diameter.

To this we ſhall add the calculation of the Area comprehended between the Hyperbola *d*FD and its Aſymptote CA, let C be the center of the Hyperbola, and putting CA $=a$, AF $=b$, and AB $=$ A*b* $=x$; it will be $\frac{ab}{a+x}=$ BD, and $\frac{ab}{a-x}=$ *bd*; whence the Area AFDB $=bx-\frac{bxx}{2a}+\frac{bx^3}{3a^2}-\frac{bx^4}{4a^3}$, &c. And the Area AF*db* $=bx+\frac{bx^2}{2a}+\frac{bx^3}{3a^2}+\frac{bx^4}{4a^3}$, &c. And the ſum *bd*DB $=2bx+\frac{2bx^3}{3a^2}+\frac{2bx^5}{5a^4}+\frac{2bx^7}{7a^6}$, &c. Now let us ſuppoſe CA $=$ AF $=1$, and A*b* or AB $=\frac{1}{10}$, C*b* being $=0 . 9$, and CB $=1 . 1$. then ſubſtituting

S 2

Newton's calculation of π in ***The Method of Fluxions.***

THE digit hunters of Newton's time returned to the Gregory series

$$\arctan x = x - x^3/3 + x^5/5 - x^7/7 + \ldots, \tag{9}$$

which they modified in various ways to accelerate its convergence. The astronomer Abraham Sharp (1651-1742), for example, substituted $x = \sqrt{(1/3)}$, which gave him

$$\frac{\pi}{6} = \frac{1}{\sqrt{3}}\left(1 - \frac{1}{3.3} + \frac{1}{3^2.5} - \frac{1}{3^3.7} + \ldots\right) \tag{10}$$

and using this series, he calculated 72 decimal places.

In 1706, John Machin (1680-1752), Professor of Astronomy in London, used the following stratagem to make the Gregory series rapidly convergent and convenient for numerical calculations as well.

For $\tan\beta = 1/5$, we have

$$\tan 2\beta = \frac{2\tan\beta}{1 - \tan^2\beta} = \frac{5}{12} \tag{11}$$

and

$$\tan 4\beta = \frac{2\tan 2\beta}{1 - \tan^2 2\beta} = \frac{120}{119} \tag{12}$$

This differs only by 1/119 from 1, whose arctangent is $\pi/4$; in terms of angles, this difference is

$$\tan(4\beta - \pi/4) = \frac{\tan 4\beta - 1}{1 + \tan 4\beta} = \frac{1}{239} \tag{13}$$

and hence

$$\begin{aligned}\arctan(1/239) &= 4\beta - \pi/4 \\ &= 4\arctan(1/5) - \pi/4\end{aligned} \tag{14}$$

Substituting the Gregory series for the two arctangents in (14), Machin obtained

$$\pi/4 = 4\left(\frac{1}{5} - \frac{1}{3.5^3} + \frac{1}{5.5^5} - \dots\right) - \left(\frac{1}{239} - \frac{1}{3.239^3} + \frac{1}{5.239^5} - \dots\right) \quad (15)$$

This was a neat little trick, for the second series converges very rapidly, and the first is well suited for decimal calculations, because successive terms diminish by a factor involving $1/5^2 = 0.04$. From (15), Machin calculated π to 100 decimal places in 1706.

The French mathematician de Lagny (1660-1734) increased the number of decimal places to 127 in 1719, but since he sweated these out by Sharp's series (10), he exhibited more computational stamina than mathematical wits.

It was also in Newton's life time that the circle ratio was first denoted by the letter π. It was used by William Jones (1675-1749), who occasionally edited and translated (from Latin) some of Newton's works, and who himself wrote on navigation and general mathematics. In 1706 he published (in English) his *Synopsis Palmariorum Matheseos: or, a New Introduction to the Mathematics.* This work was intended *for the Use of some Friends who have neither Leisure, Convenience, nor, perhaps, Patience, to search into so many different Authors, and turn over so many tedious Volumes, as is unavoidably required to make but tolerable progress in the Mathematics.* Jones' introduction of the symbol π in his book strongly suggests that he used the letter π as an abbreviation for the English word *periphery* (of a circle with unit diameter). The notation is used several times in the book, and the value calculated by Machin is reproduced; to which Jones adds *True to above a 100 places; as Computed by the accurate and Ready Pen of the Truly Ingenious Mr. John Machin.*[74]

However, Jones did not have the weight to make his notation generally accepted. The letter π was first used by Euler in his *Variae observationes circa series infinitas* (1737). Until that time, he had been using the letters *p* or *c*. Once Euler adopted it, it became a standard symbol, as was the case with his other notations.

ON March 20, 1727, Newton died. His body lay in state like that of a sovereign, and he was buried in Westminster Abbey, in a place that had often been refused to the highest nobility.

NEWTON

But not only in death did Newton's countrymen pay him the respect that he deserved. They revered this great man in his lifetime also, a marked difference from the way in which other countries treated their scientists then and in later times. In Italy, Galileo had been broken by threats, and perhaps application, of torture. France's religious intolerance drove the great mathematician Abraham De Moivre (1667-1754) into exile. Johann Kepler was debased to an astrologer by Rudolf II, who often let him go without his due salary. In the 20th century, Nazi Germany drove Einstein and hundreds of other scientists into exile and murdered Jewish scientists who had not escaped in time. And even while you are reading this, hundreds — perhaps thousands — of outstanding scientists struggle to remain alive in the horrors of Soviet forced labor camps, to which they were sentenced for disagreeing with their government, or for applying to emigrate, or even for no reason at all.

But England did not fail her greatest son.

14

EULER

He calculated just as men breathe, as eagles sustain themselves in the air.
FRANCOIS ARAGO
(1786-1853)

THE sword that Newton had forged never found a greater swordsman than the unbelievable wizard Leonhard Euler (1707-1783). If Newton was the greatest all-round scientist of all times, Euler was the greatest mathematician (in my personal estimation; many think Gauss greater).

A mere annotated index of Euler's works would fill a book far bigger than the one you are now reading; for Euler published a total of 886 books and mathematical memoirs, and his output averaged 800 printed pages a year. On the 200th anniversary of his birthday in 1907, it was decided to publish his collected works in his native country, Switzerland; by 1964, 59 volumes were published, and the entire series is expected to run to 75 volumes of about 600 pages each. And that does not include his voluminous correspondence with the Bernoullis, Goldbach, and other famous mathematicians, which contains more of his brilliant work. A conservative estimate puts the number of these letters at 4,000, of which 2,791 have been preserved; a mere list of these, with an annotation of about two lines each, fills a book of 390 pages.[75]

In 1735, Euler lost the sight of his right eye, and in 1771 he went blind completely; but for the remaining 12 years of his life, his output continued unabated, and only death cut off the avalanche of the most prolific mathematician of all times.

In our own time, much half-baked research is being fired off to scientific journals whose number, let alone contents, can no longer be followed by anyone. The middle of the 20th century has also seen the "saturation bombing" approach to scientific problems: Teams of scientists supported by batteries of computers are hired to attack a particularly important problem in the hope that something of the mass of produced material will work out. It might therefore be concluded from the above account that much of the stupendous output flowing from Euler's pen must have been second rate.

Not so. Almost all of Euler's work was brilliant, and almost all of it was of fundamental significance. True, he was stung by divergent series once or twice, and occasionally one might accuse him of a lack of rigor. But his brilliant recklessness is part of what made him so great. Euler had no time for hairsplitting, because he spent all his time in the thick of where the action was. He did not prove that one is greater than zero or that the circle has no discontinuities (as became the fashion in the 19th century), and had he done so, he might not have had the time or inclination to formulate the principle of least action or to discover the relation between exponential and trigonometric functions. All branches of mathematics abound with Euler's theorems, Euler's coefficients, Euler's methods, Euler's proofs, Euler's constant, Euler's integrals, Euler's functions, and Euler's everything else. Any modern textbook shows Euler's indelible marks in analysis, differential equations, special functions, theory of equations, number theory, differential geometry, projective geometry, probability theory, and all other branches of mathematics, and some of physics (astronomy, strength of materials, mechanics, hydrodynamics, etc.) as well. Not content with that, Euler founded, or co-founded at least two new branches of mathematics: the calculus of variations and the theory of functions of a complex variable.

From this again it might be concluded that the man was a genius on the borders of insanity, an unworldly, sinister Cyclops.

Not at all. Contemporaries describe him as a jovial fellow, witty and enjoying life. He was happily married, his household, including children and grandchildren, grew to 13 members of the family, and amidst them he would do his calculations with a child on his lap and a cat on his back. He would roll with laughter at a puppet show and indulge in horseplay with his children and grandchildren, his fantastic mind calculating away at the same time.[76]

LEONHARD EULER
(1707-1783)

Leonhard Euler was born in 1707 at Basel, Switzerland. Like Newton, he was not particularly brilliant in mathematics as a child, though he did have an advantage over Newton in that his father, a clergyman, was an amateur mathematician. In 1720, he enrolled at the University of Basel where, intending to enter the ministry, he took theological subjects, Greek, Latin and Hebrew (as a Swiss he was equally fluent in French and his native German). But as we shall see, he had a phenomenal memory, and a mere three languages left him with plenty of spare time. This he used to take physics, astronomy, medicine, and mathematics, the last taught by Jean Bernoulli I (1667-1748). He made the acquaintance of his three sons Nicolaus III (1695-1726), Daniel I (1700-1782) and Jean II (1710-1790), and through them he discovered his true vocation.

The Bernoullis are the most distinguished family in the history of mathematics. Founded by Jean I's father Nicolaus (1623-1708), this family produced a veritable stream of brilliant mathematicians, and it is still going strong in Switzerland today. It is difficult for the layman to keep the Bernoullis apart without a genealogical chart, especially since they came mostly in Jeans, Jacqueses, Nicolauses and Daniels, and the confusion of their first names is compounded by their versions in French, English, German and Latin (Jacques = James = Jakob = Jacobus; Jean = John = Johann = Ioannus). Euler formed a lifelong friendship with the three brothers named above, and in the following we shall omit the Roman numerals indentifying them in the dynasty. In 1725 Nicolaus and Daniel went to St. Petersburg, then capital of Russia, where an Academy of Sciences had been founded a few years earlier, and in 1727 Daniel arranged for Euler to obtain a somewhat humble position there as well, but after Daniel left in 1733, Euler became the Academy's chief mathematician (Nicolaus had drowned a year before Euler's arrival).

The Academy had established a scientific journal, the *Commentarii Academiae Scientiarum Imperialis Petropolitanae*, and almost from the very beginning Euler contributed to this as well as to other journals. Not only did the editors of the Petersburg Commentaries have no shortage of material as long as Euler was alive, but it took them *43 years after his death* to print the backlog of mathematical papers Euler had submitted to this journal. Asked for an explanation why his memoirs flowed so easily in such huge quantities, Euler joked that his pencil seemed to surpass him in intelligence.

In the late 1730's, Russia's government became infested with more intrigues than is ususal even for that country, and after the death of Tsarevna Anna in 1740, a series of rapid changes of power boded no good for the Academy. It was probably for this reason that Euler accepted an invitation by Frederick II of Prussia to join "my" (as Frederick used to say) Academy in Berlin.

Frederick II, surnamed by the Germans the Great, was the father of Prussian militarism. His father Frederick I had built up the army that was to be the scourge of Europe for the next 200 years, and Frederick II converted Prussia into a military camp. He was to become the idol of Josef Goebbels, who in 1944 and 1945 raved that Frederick had won the Seven Years' War even after the Russians had occupied Berlin. *Ich bin der erste Diener des Staates*, said Frederick, I am the first servant of the State. After the mediaeval Church and the Sun Kings, Frederick had discovered a new horror to be let loose on Europe, the State with a capital *S*; and to this day many a German *scheisst* in his pants on hearing the sacred word.

Euler was by this time the undisputed leader of European mathematics, and he was received with great honors at the Prussian court. At one of the receptions at Potsdam, the king's mother asked why he would not make conversation, but only answered her questions with a monosyllabic "yes" or "no." "Madam," answered Euler, "I have arrived from a country where they hang those who talk." How times have changed! In today's Russia they do them to death in forced labor camps or lunatic asylums.

Euler spent 25 years at the Prussian Academy, but he was not happy there. Frederick's interest in science was limited to warfare and the prestige of the Academy, which he lowered by using it for publishing his own writings on political and military subjects. In these, he preached Voltaire's enlightenment, while simultaneously practicing oppression. As for Euler, he ordered him about, giving him such scientific tasks as checking out the water supply in the royal palace at Potsdam.

Meanwhile in Russia, Catherine II, called the Great for somewhat similar reasons as Frederick, had clawed her way to the Russian throne. Except for her uninhibited sexual enjoyment of her many lovers, she was a rather typical forerunner of the Soviet Tsars, mercilessly centralizing her power, increasing the realm of Mother Russia by brutal conquest, paying lip service to the French enlighteners while practicing cruel oppression, and maintaining her power by all means, including the murder of her husband and later her son. She, too, wanted the prestige of having Europe's most brilliant mathematician in her Academy, to whose history he had so vigorously contributed a quarter of a century earlier, and she gave the Russian ambassador to the Prussian court instructions to meet any conditions that Euler might pose. Relations between Euler and Frederick became more and more strained, and Euler decided to accept the Russian offer. When it came to money, such unworldly men as Newton or Beethoven could drive a very hard bargain, and Euler, who was never unworldly to begin with, presented the Russian ambassador with the following conditions: He was to be Director of the Academy, with a salary of 3,000 roubles per annum; his wife to receive a pension of 1,000 roubles per annum in case of his death; three of his sons to be given good positions in St. Petersburg; and the eldest son, Johann Albrecht, to become secretary of the Academy. Mindful of Catherine's instructions, the Russian ambassador accepted, at least, one may assume, after catching his breath.

After reminding Frederick (who refused even to discuss the matter) that he was a Swiss citizen, Euler returned to Russia in 1776, never to see his native Basel again. In 1771 he lost the sight of his remaining eye, but dictating first to his children and later to a secretary specially invited from Switzerland for the purpose, the flood of his research and publications continued unabated.

Beethoven wrote the IXth symphony when he was already completely deaf; Bedrich Smetana, too, wrote the cycle of symphonic poems *My Country* in utter deafness. Euler, totally blind, wrote many works such as his famous treatise on celestial mechanics, *Theoria motus planetarum et cometarum*, in which he tackled the three-body problem and for the first time introduced as the center of planetary motion not the sun, but the center of mass of the sun and the corresponding planet. How can a blind man write mathematical treatises that are none too easy to follow for us seeing mortals?

Euler had a phenomenal memory. Plagued by insomnia one night, he calculated the 6th powers of the first 100 integers in his head, and several days later he could still remember the entire table.

His uncanny ability for numerical calculations is illustrated by his refutation of Fermat's conjecture that all numbers of the form 2 to the power 2^n plus one are primes.

Not at all, said Euler in 1732; 2 to the power 2^5 plus one makes 4,294,967,297, and that is factorable into 6,700,417 times 641.

AFTER all this, it is surely not surprising that Euler delivered formulas for π by the truckload. And the truckload was usually quite casually attached to a trainload of more important items. Take, for example, one of the many formulas Euler derived for the square of π.

The series of inverse squares

$$1/1^2 + 1/2^2 + 1/3^2 + \dots \qquad (1)$$

had baffled mathematicians for decades. Gottfried Wilhelm Leibniz (1646-1716), co-inventor of the calculus, had been unable to sum it, and so had many lesser mathematicians. Jacques Bernoulli I (1654-1705), uncle of the three brothers whom Euler had befriended, proved the convergence of the series, but could not sum it either; his brother Jean I wrote that in the end he "confessed that all his zeal had been mocked." But Euler, in 1736, solved the problem in his stride. The series

$$\sin x = x - x^3/3! + x^5/5! - x^7/7! + \dots \qquad (2)$$

was already known to Newton. Euler substituted $x^2 = y$ and regarded the equation

$$\sin x = 0 \qquad (3)$$

as an equation of degree infinity, obtaining (for $y \neq 0$)

$$1 - y/3! + y^2/5! - y^3/7! + \dots = 0. \qquad (4)$$

But the roots of the equation (3) are 0, $\pm\pi$, $\pm 2\pi$, $\pm 3\pi$, ..., and therefore the roots of (4) are π^2, $(2\pi)^2$, $(3\pi)^2$, ... (zero has been excluded above). Now Euler knew from the theory of equations, a branch of higher algebra that is also heavily marked by his footprints, that the negative coefficient of the linear term [+1/3! in (4)] is the sum of the reciprocal roots of the equation, and hence

$$1/\pi^2 + 1/(2\pi)^2 + 1/(3\pi)^2 + \ldots = 1/3!,$$

or

$$1/1^2 + 1/2^2 + 1/3^2 + \ldots = \pi^2/6, \tag{5}$$

which solves the problem that had baffled all the others, and gives a series for π^2 into the bargain. But there is no stopping Euler now. Repeating the procedure for the cosine series, he finds

$$\pi^2/8 = 1/1^2 + 1/3^2 + 1/5^2 + \ldots \tag{6}$$

and subtracting (5) from twice the sum (6), he obtains

$$\pi^2/12 = 1/1^2 - 1/2^2 + 1/3^2 - 1/4^2 + \ldots \tag{7}$$

Generalizing the summation to any even power of reciprocals, i.e., considering the series of terms $1/j^i$ $(j = 1, 2, 3, \ldots)$, Euler found a general formula involving Bernoulli numbers and gave special cases for

$$1/1^4 + 1/2^4 + 1/3^4 + \ldots = \frac{2^2}{5!\,3}\pi^4 \tag{8}$$

down to

$$1/1^{26} + 1/2^{26} + 1/3^{26} + \ldots = \frac{2^{24} . 76977927}{27!}\pi^{26} \tag{9}$$

These results are given in Euler's famous textbook *Introductio in Analysin infinitorum* (1748), the book that standardized mathematical notation almost as we use it today. The symbols π, e, i, Σ, $\int$, $f(x)$ and many others are all souvenirs of Euler (though not necessarily of the *Introductio*).

To calculate the logarithm of π, Euler found infinite products for even powers of π, e.g.,

$$\frac{\pi^2}{6} = \frac{2^2}{2^2 - 1} \cdot \frac{3^2}{3^2 - 1} \cdot \frac{5^2}{5^2 - 1} \cdot \frac{7^2}{7^2 - 1} \cdots \tag{10}$$

This is not all, but it will do as a sample of this particular truckload.

Machin's neat little trick with the arctangents (pp. 144-5) turned out to be a pebble of another Eulerian truckload. Euler derived the formulas

$$\arctan\frac{1}{p} = \arctan\frac{1}{p+q} + \arctan\frac{q}{p^2+pq+1} \tag{11}$$

and

$$\arctan\frac{x}{y} = \arctan\frac{ax-y}{ay+x} + \arctan\frac{b-a}{ab+1} + \arctan\frac{c-b}{cb+1} + \ldots \tag{12}$$

and this gives rise to any amount of relations for π; for example, if the odd numbers are substituted for $a, b, c, \ldots$, we obtain

$$\pi/4 = \arctan(1/2) + \arctan(1/8) + \arctan(1/18) + \ldots \tag{13}$$

All of these formulas depend on a series for the arctangent, and Euler found one that converged more quickly than any other:

$$\arctan x = (y/x)\left(1 + \frac{2}{3}y + \frac{2.4}{3.5}y^2 + \frac{2.4.6}{3.5.7}y^3 + \ldots\right) \tag{14}$$

where

$$y = x^2/(1+x^2) \tag{15}$$

Using Machin's stratagem in the form

$$\pi = 20\arctan(1/7) + 8\arctan(3/79) \tag{16}$$

and evaluating these two terms by (14), Euler calculated π to 20 decimal places in one hour!

THESE are but a few examples of the many expressions that Euler found for π; they also included infinite products and continued fractions. So thoroughly did Euler deal with the problems (inciden-

138. Ponatur denuo in formulis § 133 arcus z infinite parvus et sit n numerus infinite magnus i, ut iz obtineat valorem finitum v. Erit ergo $nz = v$ et $z = \frac{v}{i}$, unde sin. $z = \frac{v}{i}$ et cos. $z = 1$; his substitutis fit

$$\cos. v = \frac{\left(1 + \frac{v\sqrt{-1}}{i}\right)^i + \left(1 - \frac{v\sqrt{-1}}{i}\right)^i}{2}$$

atque

$$\sin. v = \frac{\left(1 + \frac{v\sqrt{-1}}{i}\right)^i - \left(1 - \frac{v\sqrt{-1}}{i}\right)^i}{2\sqrt{-1}}.$$

In capite autem praecedente vidimus esse

$$\left(1 + \frac{z}{i}\right)^i = e^z$$

denotante e basin logarithmorum hyperbolicorum; scripto ergo pro z partim $+v\sqrt{-1}$ partim $-v\sqrt{-1}$ erit

$$\cos. v = \frac{e^{+v\sqrt{-1}} + e^{-v\sqrt{-1}}}{2}$$

et

$$\sin. v = \frac{e^{+v\sqrt{-1}} - e^{-v\sqrt{-1}}}{2\sqrt{-1}}.$$

Ex quibus intelligitur, quomodo quantitates exponentiales imaginariae ad sinus et cosinus arcuum realium reducantur. Erit vero

$$e^{+v\sqrt{-1}} = \cos. v + \sqrt{-1} \cdot \sin. v$$

et

$$e^{-v\sqrt{-1}} = \cos. v - \sqrt{-1} \cdot \sin. v.$$

Euler's Theorem as first stated in Vol. 1, Chapt. 7, Section 138 of the *Introductio* (1748). At that time, Euler was still using i to denote infinity; only later did he introduce ∞ for infinity, and i for the square root of -1.

tally) associated with π that no one after him ever found a better way of calculating its value, and he may be said to have finished off its history as far as numerical evaluation is concerned. A few more continued fractions were found after Euler by Lambert and others, and Laplace found an ingenious new approach which we shall consider in the next chapter; but none of these yields the numerical value of π faster than Euler's method (which includes Machin's series as a special case).

But if Euler finished off one chapter in the history of π, he also started another. What kind of number was π? Rational or irrational? With each new decimal digit the hope that it might be rational faded, for no period could be found in the digits. There was no proof as yet, but most investigators sensed that it was irrational. However, Euler asked a new question: Could π be the root of an algebraic equation of finite degree with rational coefficients? By merely asking the question, Euler opened a new chapter in the history of π, and a very important one, as we shall see. He was also the one who started writing it, for later investigations were based on one of Euler's greatest discoveries, the connection between exponential and trigonometric functions,

$$e^{ix} = \cos x + i \sin x \tag{16}$$

Euler discovered a long, long list of theorems. They are known as "Euler's theorem on..." and "Euler's theorem of..." But this one is simply known as *Euler's Theorem.*

Euler also laid the foundations for the investigation of the irrationality, and later the transcendence, of π by deriving the continued fractions

$$\tfrac{1}{2}(e-1) = \cfrac{1}{1 + \cfrac{1}{6 + \cfrac{1}{10 + \cfrac{1}{14 + \cfrac{1}{18 + \cfrac{1}{22 + \dots}}}}}} \tag{17}$$

$$\tanh x = \cfrac{1}{1/x + \cfrac{1}{3/x + \cfrac{1}{5/x + \cfrac{1}{7/x + \cfrac{1}{9/x + \dots}}}}} \tag{18}$$

$$\tanh(x/2) = \cfrac{1}{2/x + \cfrac{1}{6/x + \cfrac{1}{10/x + \cfrac{1}{14/x + \dots}}}} \tag{19}$$

which later formed the starting point of Lambert's and Legendre's investigations.

All of these were already contained in Euler's *Introductio in analysin infinitorum* in 1748. In 1755, already blind, Euler wrote a treatise entitled *De relatione inter ternas pluresve quantitates instituenda*, which was published ten years later (such was the backlog of papers showered on the Petersburg editors by the blind genius). Here he wrote "It appears to be fairly certain that the periphery of a circle constitutes such a peculiar kind of transcendental quantities that it can in no way be compared with other quantities, either roots or other transcendentals." (*Unde sententia satis certa videtur, quod peripheria circuli tam peculiare genus quantitatum transcendentium constituat, ut cum nullis aliis quantitatibus, sive surdis sive alius generis transcendentibus nullo modo se comparari patiatur.*)

As always, Euler was right. But it took another 107 years to prove his conjecture.

And with this we leave the old wizard. Laplace is one of the next whom we shall meet; it was he who told his students, *Lisez Euler, lisez Euler, c'est notre maître à tous.* Read Euler, read Euler, he is our master in everything.

15

The Monte Carlo Method

Tremblez, ennemis de la France,
Rois ivres de sang et d'orgeuil!
Le peuple souverain s'avance,
Tyrans, descendez au cerceuil! [77]
French revolutionary song
of Lazare Carnot's time.

PROBABILITY theory is the mathematics of the 20th century. Its history goes back to the 16th century, but not until the present century did physicists and engineers fully realize that nature and the real world can be described exhaustively only by the laws governing their randomness. What physicists had considered exact until relatively recently, turned out to be merely the mean value of a much more impressive structure; and mean values can be very misleading. ("Put one foot in an ice bucket, and the other in boiling water; then on the average you will be comfortable.") Strange to relate, even as brilliant and recent a physicist as Albert Einstein regarded the probabilistic laws of quantum mechanics as testimony to our ignorance rather than as a valid description of the laws of nature.

The beginnings of probability theory go back to the *Liber de ludo aleae* (The book of games of chance), written about 1526 by Gerolamo Cardano (1501-1576), though not published until 1663. Cardano, of cubic equation fame (p. 90), was not only a mathematician, engineer and physician, but also a passionate gambler. Until the advent of the kinetic theory of gases in the 19th century, probability theory was rarely applied to anything else but gambling. The main contributors to its development were Jacques Bernoulli I (1654-1705, author of *Ars conjectandi*), Pascal, De Moivre, Euler, Laplace, Gauss and Poisson (1781-1840), followed by a large number of mathematicians in the 19th and 20th centuries.

The number π appears in probability theory very frequently, as it does in all branches of higher mathematics; but nowhere is its appearance more fascinating than in a problem posed and solved by George Louis Leclerc, Comte de Buffon (1707-1788). Buffon (as everybody calls him) was an able mathematician and general scientist, who shocked the world by estimating the age of the earth to be about 75,000 years, although every educated person in the 18th century knew that it was no older than about 6,000 years. Among his exploits is a test of one of Archimedes' supposed engines of war used in the defense of Syracuse. As told by Plutarch, the story includes a plausible description of the action of Archimedes' cranes and missile throwers, but by the Middle Ages, it had grown into a much exaggerated legend, and the *Book of Histories* by the Byzantine author John Tzetzes (ca. 1120-1183) repeats the story with many embellishments, such as the statement that Archimedes had burned the Roman ships to ashes at a distance of a bow shot by focusing the sun's beams onto the Roman fleet. The story (which is not contained in Plutarch's description) has persisted in many books down to our own days. Buffon, a man of considerable means and spare time, decided to test the feasibility of such a machine. Using 168 flat mirrors six by eight inches in an adjustable framework, he was able to ignite wooden planks at a distance of 150 feet, and he satisfied himself that Archimedes' alleged exploit was feasible. He did not, however, satisfy posterity, since the Syracusans would hardly have had the same leisure to focus 168 beams, nor would the Roman ships floating on the sea have held as still as Buffon's beams on the ground.

But back to Buffon's problem involving π. The problem which he posed (and solved) in 1777 was the following: Let a needle of length L be thrown at random onto a horizontal plane ruled with parallel straight lines spaced by a distance d (greater than L) from each other. What is the probability that the needle will intersect one of these lines?

We assume that "at random" means that any position (of the center) and any orientation of the needle are equally probable and that these two random variables are independent. Let the distance of

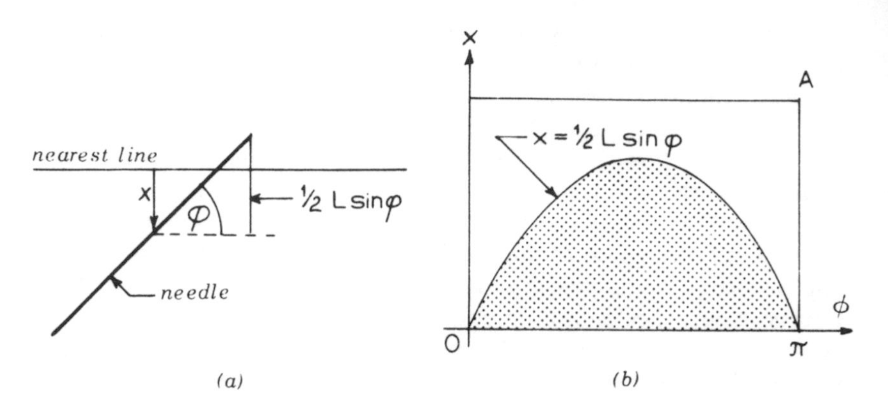

Buffon's problem.

the center of the needle from the nearest line be x, and let its orientation be given by ϕ (figure *a*). Since x is measured from the *nearest* line, we need only consider a single line, because the others involve only repetition of the same solution.

It is obvious from the figure that the needle will intersect a line if and only if

$$x < \tfrac{1}{2} L \sin\phi . \tag{1}$$

The problem is therefore equivalent to finding the probability

$$P(x < \tfrac{1}{2} L \sin\phi).$$

To find this probability, use the plane of rectangular coordinates x, ϕ, and consider the interior of the rectangle OA (figure *b*) whose points satisfy the inequalities

$$\begin{aligned} 0 &< x < d/2 \\ 0 &< \phi < \pi \end{aligned} \tag{3}$$

These are the intervals of possible values of x and ϕ, and therefore any point inside the rectangle OA corresponds to one and only one possible combination of position (x) and orientation (ϕ) of the needle. Since all such combinations are equiprobable, and the area of the rectangle represents the sum total of all possibilities that can arise (because, not quite beyond reproach, we regard this area as made up of all points inside it). However, not all of these possibilities will result in an intersection of the needle with a line; such an intersection, as we have found, will take place only under the condition (1), that is, for positions and orientations corresponding to points lying below the curve $x = \frac{1}{2}L \sin\phi$ in figure *b*, so that the sum total of possibilities resulting in the intersection by the needle is given by the area under this curve. If, then, probability is the ratio of the number of favorable, to the number of possible, events under given conditions, the probability of intersection is given by the ratio of the shaded part to the entire rectangle *OA* in figure *b*, that is, the required probability (2) is

$$P = \tfrac{1}{2}L \int_0^{\pi} \sin\phi \, d\phi \ : \ \pi d/2 = 2L/\pi d \qquad (4)$$

This is the result Buffon derived. He also attempted an experimental verification of his result by throwing a needle many times onto ruled paper and observing the fraction of intersections out of all throws. Whether he modified his result for an evaluation of π I do not know, but the problem and its solution were largely forgotten for the next 35 years, until one of the great mathematicians with whom France has been blessed, called attention to it and gave it a new twist.

Pierre Simon Laplace (1749-1827) was one of the "three great L's" among French mathematicians of the time. The other two, Joseph Louis Lagrange (1736-1813) and Adrien Marie Legendre (1752-1833), were his contemporaries, and all three survived the French Revolution as members of the Committee of Weights and Measures, which discarded the cubits, feet, pounds and miles of the old regime and worked out the metric system as we use it today. It was, incidentally, another mathematician, Lazare Carnot (1753-1823), who saved the young French republic in its hour of greatest need. Scared out of their wits by the cry for liberty, equality and fraternity, Europe's kings, princes, princelings, counts and whatnots turned on the Revolution. Threatened by internal confusion and the invading armies deep inside France, the Revolution seemed about to be crushed; but Carnot, member of the Committee for Public Safety in charge of military affairs, took command and sent the invaders packing on all fronts, becoming *organisateur de la victoire*, the hero of the French Revolution. But like so many other sincere revo-

JOSEPH LOUIS LAGRANGE
(1736-1813)

PIERRE SIMON LAPLACE
(1749-1827)

lutionaries after him, Carnot soon observed that a revolution only replaces one tyranny by another, and refusing to go along with its excesses, he was driven into exile as a "royalist." Significantly, his chair of geometry at the *Institut National* was unanimously voted to a general; a general by the name of Napoleon Bonaparte, another one in a long line of power-hungry careerists who was to preach liberty and practice oppression.

PIERRE Simon Laplace is known, above all, for authoring two masterpieces, *Mecanique céleste* (5 vols., 1799-1825) and *Théorie analytique des probabilites* (1812). The former was the greatest work on celestial mechanics since Newton's *Principia*, including many new mathematical techniques, such as the theory of potential. Asked by Napoleon why in the entire work on celestial mechanics he had not once mentioned God, Laplace replied, *Sire, je n'avais pas besoin de cette hypothèse* — Sire, I had no need of that hypothesis. Napoleon, incidentally, appointed Laplace Minister of Interior, but after six weeks dismissed him again, commenting that he "carried the spirit of the infinitely small into the management of affairs." The *Théorie analytique* is the foundation of modern probability theory. Among many new mathematical techniques it contains the integral transform that is today the daily bread of every systems engineer and analyst of electrical circuits.

It also contains a discussion of Buffon's problem, which Laplace saw in a new light. From the first and last expressions in (4) we have

$$\pi = 2L/dP \tag{5}$$

and this is an entirely new method of evaluating π: The length of the needle L and the spacing between the lines d are known (usually one makes $L = d$), and the probability of intersection P can be measured by throwing a needle onto ruled paper a very large number of times, recording the fraction of throws resulting in an intersection of the needle with a line.

This method, which Laplace generalized for paper with two sets of mutually perpendicular lines, has been used by several people as a playful diversion to calculate the first decimal places of π by thousands of throws. One of them was a certain Captain Fox, who indulged in this sport while recovering from wounds incurred in the American Civil War.[78]

It is not difficult to calculate the probability of obtaining π correct to k decimal places in N throws.[79] The results of such a calculation show that this method is very inefficient as far as the numerical computation of π is concerned; for example, the probability of obtaining π correct to 5 decimal places in 3,400 throws of the needle is less than 1.5%, which is very poor.

Nevertheless, Laplace had discovered a powerful method of computation that did not come into its own until the advent of the electronic computer. The method that Laplace proposed consists in finding a numerical value by realizing a random event many times and observing its outcome experimentally. This is today known as a Monte Carlo method (Monte Carlo is the European Las Vegas), and it is used in a wide field of applications ranging from economics to nuclear physics.

Let us first take the example of calculating π by this method. A computer can easily throw a needle 500 times a second, or 1.8 million times per hour. Not literally, of course, but it can be programmed to select a random number (x) for the position of the needle and another (ϕ) for its orientation, which is just as good; it *simulates* the throwing of a needle. It can also be programmed to observe whether the needle has intersected or not, that is, whether the inequality (1) is satisfied or not. Finally, it is programmed to record the number of intersections in the total number of throws, and after computing the resulting value of π by formula (5), to print the value it has found.

A program of this type is shown on the next page. It is written in BASIC, a simple, but powerful computer language.

```
 10 LET N = 0
 20 PRINT 'NO. OF THROWS', 'PI'
 30 FOR J = 1 TO 24
 40 FOR K = 1 TO 500
 50 LET X = RND(X)
 60 LET U = SIN(3.1415927*RND(F))
 70 IF X GT U THEN 90
 80 LET N = N + 1
 90 NEXT K
100 LET T = 500*J
110 LET P = 2*T/N
120 PRINT T, P
130 NEXT J
140 END
```

The program was actually run[80] and resulted in the following values of π in the first 12,000 throws:

NO. OF THROWS	PI
500	3.2154341
1000	3.2414911
1500	3.1645569
2000	3.1620553
2500	3.1407035
3000	3.1430068
3500	3.1460674
4000	3.1421838
4500	3.1435557
5000	3.1446541
5500	3.1401656
6000	3.1217482
6500	3.1175060
7000	3.1354983
7500	3.1453135
8000	3.1452722
8500	3.1493145
9000	3.1441048
9500	3.1384209
10000	3.1392246
10500	3.1493701
11000	3.1527658
11500	3.1485284
12000	3.1417725

We cannot in this way, of course, obtain a better value of π than the one we inserted in line 60 of the program on the opposite page (it occurs there to make the orientation of the needle uniformly distributed between 0° and 180°), and the same line results in an error owing to a series of successive roundings off. However, even if these technicalities were corrected, the result would still be poor, as predicted by the calculation of the probability of obtaining k correct decimal places in a series of n throws. In the same processing time (53 seconds) we could have obtained a much better value, for example, by Euler's method.

But if the method is not very efficient for calculating π, it is very powerful in other applications. Suppose, for example, that we wish to calculate the mean value of a complicated function of a random variable. This is found by an integration involving the probability density function of the random variable. But sometimes the resulting integral is so complicated that it takes a long time to write the program and that it involves a costly amount of processing time. In that case we do not program the computer for the complicated evaluation of the integral, but we make it simulate the random variable and its function and we make it compute the arithmetic mean of, say, one hundred thousand trials. The result is the required mean value.

Or suppose we wish to find a complicated multiple integral. A Monte Carlo method of finding it (instead of writing a cumbersome program) is to let the computer "shoot" n-tuplets of random numbers. These represent a coordinate in (n-dimensional) space and the coordinate either lies in the volume determined by the integral ("hit") or it does not ("miss"). Then we let the computer shoot at the target, say, half a million times. The number of hits is then proportional to the n-tuple integral.

The man who taught us to program electronic computers in this way was Pierre Simon Laplace. His computer was neither electronic nor digital. It was an analog computer consisting of one needle and one piece of ruled paper.

16

The Transcendence of π

Frustra laborant quotquot se calculationibus fatigant pro inventione quadraturae circuli.

Futile is the labor of those who fatigue themselves with calculations to square the circle.

MICHAEL STIFEL (1544)

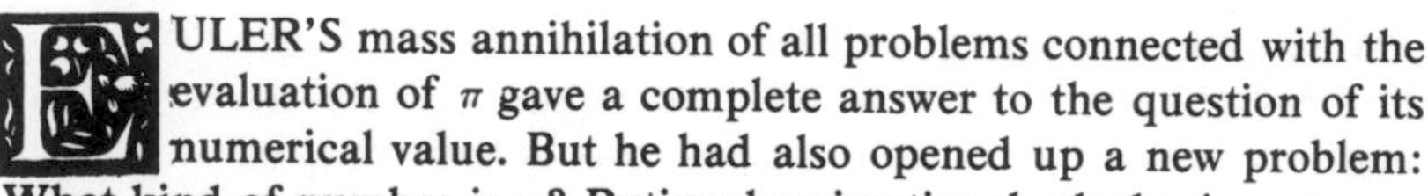

EULER'S mass annihilation of all problems connected with the evaluation of π gave a complete answer to the question of its numerical value. But he had also opened up a new problem: What kind of number is π? Rational or irrational, algebraic or transcendental? That question was to haunt mathematicians for another 107 years after Euler had asked it.

Already the Greeks before Euclid's time were familiar with the existence of irrational numbers, that is, of numbers that could not be expressed as ratios of two integers. They expressed the concept in different words, since the only branch of mathematics with which they were thoroughly familiar was geometry; they said (for example) that the diagonal of a square was *incommensurable* with its side, the ratio of the two being $\sqrt{2} : 1$. Even Aristotle, though shockingly ignorant of the science of his time, was dimly aware of the proof that $\sqrt{2}$ is irrational. The proof was a *reductio ad absurdum* and ran as follows:

Let $\sqrt{2}$, if possible, be equal to a fraction p/q, where the integers p and q are in their lowest terms. Then not more than one of the two integers can be even (or we could cancel by 2). Since $2q^2 = p^2$, it follows that p^2, and hence p, must be even ($p = 2r$), so that q must be odd. But from $2q^2 = 4r^2$ we find q^2, and hence q, even; therefore q is both even and odd, which is absurd; hence the assumption $\sqrt{2} = p/q$ is false.

There is, however, no reason why an irrational number should not be a root of an algenraic equation; for example, $\sqrt{2}$ is a solution of the algebraic equation $x^2 - 2 = 0$. An algebraic equation is an equation

$$a_n x^n + \ldots + a_2 x^2 + a_1 x + a_0 = 0 \tag{1}$$

where n is finite and all coefficients a_i are rational (or, which amounts to the same thing, integers, because if they are rational, we can always multiply the equation by a common multiple of the denominators).

By Euler's time, people began to suspect that there were "worse" numbers than irrational ones, namely numbers that were not only irrational, but that could not even be roots of an algebraic equation. Such numbers are called *transcendental.*

It was not at all obvious that such numbers exist. For example, the equation

$$\sin x = \tfrac{1}{2} \tag{2}$$

is transcendental, because it is not algebraic; if we expand the sine in a power series, we have

$$-\tfrac{1}{2} + x - x^3/3! + x^5/5! + \ldots = 0 \tag{3}$$

and this is not like (1) because it violates the condition that n must be finite. One of its solutions is $x = \pi/6$; but who says that a transcendental equation must have a transcendental solution? The equation $\sin x = 0$ is also transcendental, but one of its solutions, $x = 0$, is obviously an algebraic number.

The existence of transcendental numbers was not proved until 1840, when Joseph Liouville (1809-1882) showed that one could define numbers (in his proof, as limits of continued fractions) which cannot be roots of any algebraic equation.

But granted that transcendental numbers exist, why should they be of interest? The answer is that they have many interesting properties, and more specifically, that the transcendence of π provides an immediate answer to the age-old problem of whether the circle is squarable.

One of the interesting properties of transcendental numbers is that there are "more" of them than there are algebraic numbers. The "more" is in quotation marks because the set of algebraic numbers is infinite, just as the set of transcendental numbers is infinite. However, there are ways to compare one infinite set to another. When we count a number of objects, say 17 trees in a garden, we are really comparing their amount to the set of natural numbers 1, 2, 3, . . . , 17. By counting the trees, we are establishing a one-to-one correspondence between each tree and each natural number. We stop the process when this correspondence can no longer be established, in this case for numbers larger than 17, with which no trees can be associated. We then know that the amount of trees in the garden is as large as the amount of numbers in the set of the first 17 natural numbers. But this process of counting can be extended to infinity; any set whose members can be brought into a one-to-one correspondence with the members of the set of natural numbers can be counted this way, even if the counting goes on forever. Such a (countable) set is said to be *denumerable*, even though it may be infinite. The set of even numbers, for example, is evidently denumerable, so that it is "just as big" as the set of *all* natural numbers, even or odd. One might have expected it to be only "half as big" as the set of natural numbers, but that is an impermissible generalization of what we are used to in finite sets. The founder of set theory, Georg Cantor (1845-1918), showed that the set of rational, and even of algebraic, numbers is denumerable, but that the set of transcendental (and hence also real) numbers is non-denumerable. These infinites of different orders can be described by new numbers, called cardinal numbers, and these transfinite numbers have their own arithmetic, showing that the joys of mathematics are truly endless: They go beyond infinity.

But the reason why the transcendence of π comes into our story is quite different. We have already seen the reasons why the Greeks insisted that the circle be squared only by a finite number of constructions using compasses and straightedge only (pp. 50-51). After Descartes had found the new geometry, there was the possibility of determining the feasibility of such a construction analytically. A circle can be squared if it can be rectified; if its diameter is unity, we must construct a line of length π. By using compasses and ruler only, we can draw only straight lines and circles, that is, curves whose equations are polynomials of not more than second degree. The points obtained by successive constructions are therefore always intersections (or tangent points) of curves of not more than second degree. We are given a circle whose equation is (we assume unit diameter)

$$4x^2 + 4y^2 = 1 \tag{4}$$

and the final result of the construction is to be a distance equal to π. The coordinates of the end point of this distance is obtained by a chain of constructions, each of which amounts to the following: We are given certain points (from the previous construction) whose coordinates are known numbers; these coordinates (or their simple functions) become the coefficients of the equation that is to be solved in the next step, since an intersection involves the solution of two simultaneous equations.

Starting with (4) and the next step in the construction, characterized by a curve of not more than second degree, we find the intersection of the two curves by solving at most a quadratic equation, whose roots are either rational or irrational involving only square roots. These roots, or their simple functions, become the coefficients of the equation to be solved in the next step of the construction. The next equation is therefore quadratic with coefficients that are either rational or square roots. To convert this to an equation with rational coefficients, it is sufficient (and not even necessary) to square the equation twice over, resulting in an equation of not more than 8th degree. If the construction has s steps, the final equation to be solved in order to yield the length π must therefore be an equation with coefficients of degree not higher than 8^s, where s is to be finite.[81] If the rectification (or squaring) of the circle is possible in a finite number of quadratic steps, then one of the roots of this algebraic equation is π (or $\sqrt{\pi}$); but if π is a number that is not the root of any algebraic equation, then the rectification and squaring of the circle (by Greek rules) is impossible.

Thus, the question of whether the circle can be squared by Euclidean geometry could be answered as follows: *If* π *is a transcendental number, then it cannot be done.*

The theory of equations thus had, as modern jargon goes, "put a handle on the problem" which Anaxagoras had pondered on the floor of his prison in Athens in the 5th century B.C. When Lindemann in 1882 finally proved that π is transcendental, he finished off the prey with the last blow; but it was the above line of reasoning that first brought it to bay. Although it cannot be attributed to any single man, the mathematician to whom the theory of equations owes more than to anybody else was Carl Friedrich Gauss (1777-1855).

Gauss also estimated the value of π by using lattice theory and considering a lattice inside a large circle,[83] but this is as close as this great mathematician came to the story of π.

LONG before Liouville's proof of the existence of transcendental numbers (1840), the irrationality of π was established by the Swiss mathematician Johann Heinrich Lambert (1728-1777), and by Adrien-Marie Legendre (1752-1833). Lambert proved the irrationality of π in 1767, but Legendre provided a more rigorous proof of an auxiliary theorem concerning continued fractions that Lambert had used. In his treatise *Vorläufige Kenntnisse für die, so die Quadratur und Rectification des Circuls suchen* (Preliminary knowledge for those who seek the quadrature and rectification of the circle, 1767), Lambert investigated certain continued fractions and proved the following theorem:

If x is a rational number other than zero, then tan *x cannot be rational.*

From this it immediately follows that

If tan *x is rational, then x must be irrational or zero* (for if it were not so, the original theorem would be contradicted).

Since $\tan(\pi/4) = 1$ is rational, $\pi/4$ must be irrational, and hence the irrationality of π is established.

Lambert also gave an interesting continued fraction for π,

$$\pi = 3 + \cfrac{1}{7 + \cfrac{1}{15 + \cfrac{1}{1 + \cfrac{1}{292 + \cfrac{1}{1 + \cfrac{1}{1 + \cfrac{1}{1 + \cfrac{1}{2 + \dots}}}}}}}}$$

The *inverse* convergents of this continued fraction are shown on the opposite page. The first convergents of Lambert's continued fraction had been found in some way or other long before his time:

3	value implied by I Kings vii, 23;
22/7	upper bound found by Archimedes, 3rd century B.C.;
333/106	lower bound found by Adriaan Anthoniszoon, ca. 1583;
355/113	found by Valentinus Otho, 1573;[84] also Anthoniszoon, Metius and Viète (all 16th century).

1 : 3
7 : 22
106 : 333
113 : 355
33102 : 103993
33215 : 104348
66317 : 208341
99532 : 312689
265381 : 833719
364913 : 1146408
1360120 : 4272943
1725033 : 5419351
25510582 : 80143857
52746197 : 165707065
78256779 : 245850922
131002976 : 411557987
340262731 : 1068966896
811528438 : 2549491779
1963319607 : 6167950454
4738167652 : 14885392687
6701487259 : 21053343141
567663097408 : 1783366216531
1142027682075 : 3587785776203
1709690779483 : 5371151992734
2851718461558 : 8958937768937
107223273857129 : 336851849443403
324521540032945 : 1019514486099146 etc.

Inverse convergents given by Lambert.

Legendre, in his *Elements de Géometrie* (1794) proved the irrationality of π more rigorously, and also gave a proof that π^2 is irrational, dashing the hopes that π might be the square root of a rational number. Toward the end of his investigation, Legendre writes: "It is probable that the number π is not even contained among the algebraic irrationalities, i.e., that it cannot be the root of an algebraic equation with a finite number of terms, whose coefficients are rational. But it seems to be very difficult to prove this strictly."

Legendre was right on both counts: π is not algebraic, and the proof so difficult that it was not found for 88 more years.

In 1873, Charles Hermite (1822-1901) proved that the number e is transcendental; from this it follows that the finite equation

$$ae^r + be^s + ce^t + \dots = 0 \tag{6}$$

cannot be satisfied if $r, s, t \dots$ are natural numbers and $a, b, c, \dots$ are rational numbers not all equal to zero.

In 1882, F. Lindemann[85] finally succeeded in extending Hermite's theorem to the case when $r, s, t, \dots$ and $a, b, c, \dots$ are algebraic numbers, not necessarily real. Lindemann's theorem can therefore be stated as follows:

If $r, s, t, \dots, z$, are distinct real or complex algebraic numbers, and $a, b, c, \dots, n$ are real or complex algebraic numbers, at least one of which differs from zero, then the finite sum

$$ae^r + be^s + ce^t + \dots + ne^z \tag{7}$$

cannot equal zero.

From this the transcendence of π follows quickly. Using Euler's Theorem in the form

$$e^{i\pi} + 1 = 0, \tag{8}$$

we have an expression of the form (7) with $a = b = 1$ algebraic, and c and all further coefficients equal to zero; $s = 0$ is algebraic, leaving $r = i\pi$ as the only cause why (8) should vanish. Thus, $i\pi$ must be transcendental, and since i is algebraic, π must be transcendental.

It stands to reason that a proof which was 100 years in the making is neither short nor easy. Lindemann's paper[85] runs to 13 pages of tough mathematics. Karl Wilhelm Weierstrass (1815-1897), the apostle of mathematical rigor, simplified the proof of Lindemann's theorem somewhat in 1885,[86] and it was further simplified in later years by renowned mathematicians (Stieltjes, Hurwitz, Hilbert and others). The interested reader is referred to the comparatively easy version given by Hobson.[87]

As we have seen, the possibility of squaring the circle by Euclidean construction hinged entirely on the question whether π was algebraic or transcendental; Lindemann's theorem therefore proved that the squaring of the circle by the rules of Greek geometry is impossible.

And that is the end of the history of π and squaring the circle. Or it would have been if there were no fools among us.

These fools rushed in where sages had trodden for 2,500 years.

17

The Modern Circle-Squarers

This invention relates to a device which renders it impossible for the user to stand upon the privy-seat; and consists in the provision of rollers on top of the seat, which, although affording a secure and convenient seat, yet, in the event of an attempt to stand upon them, will revolve, and precipitate the user on to the floor.
U.S. PATENT No. 90,298 (1869).

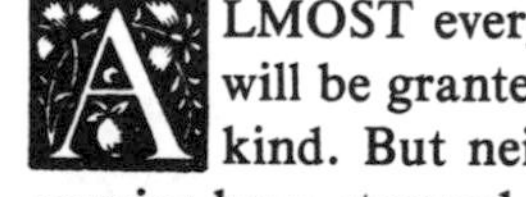LMOST every country now has a law stating that no patent will be granted for an invention of a *perpetuum mobile* of any kind. But neither man-made laws nor the laws of thermodynamics have stopped an army of mavericks from designing hundreds of versions of an alleged *perpetuum mobile*.

Similarly, in 1775 the *Académie Française* passed a resolution henceforth not to examine any more solutions of the problem of squaring the circle. But undaunted by either the Academy's resolution or Lindemann's proof, the circle squarers marched on; and they are still marching, spiteful of the cruel world that will not recognize their grand intellectual achievements. "The race of circle squarers," says Schubert,[98] "will never die out as long as ignorance and the thirst for glory remain united."

Every country has its circle-squarers, but the following will be limited to the circle-squarers of America.

There is a story about some American legislature having considered a bill to legislate, for religious reasons, the biblical value of $\pi = 3$. I have found no confirmation of this story; very probably it grew out of an episode that actually took place in the State Legislature of Indiana in 1897. The Indiana House of Representatives did consider and unanimously pass a bill that attempted to legislate the value of π (a wrong value); the author of the bill claimed to have squared the circle, and offered this contribution as a free gift for the sole use of the State of Indiana (the others would evidently have to pay royalties).

The author of the bill was a physician, Edwin J. Goodman, M.D., of Solitude, Posey County, Indiana, and it was introduced in the Indiana House on January 18, 1897, by Mr. Taylor I. Record, Representative from Posey County. It was entitled "A bill introducing a new Mathematical truth,"[89] and it became House Bill No. 246; copies of the bill are preserved in the Archives Division of the Indiana State Library; the full text has also been reprinted in an article by W.E. Eddington in 1935.[90]

The preamble to the bill informs us that this is

> *A bill for an act introducing a new mathematical truth and offered as a contribution to education to be used only by the State of Indiana free of cost by paying any royalties whatever on the same, provided it is accepted and adopted by the official action of the legislature in 1897.*

The bill consisted of three sections. Section 1 starts off like this:

> *Be it enacted by the General Assembly of the State of Indiana: It has been found that the circular area is to the quadrant of the circumference, as the area of an equilateral rectangle is to the square on one side. The diameter employed as the linear unit according to the present rule in computing the circle's area is entirely wrong...*

An "equilateral rectangle" is, of course, a square, so that the first statement does not make any sense at all; but if we give the author the benefit of the doubt and assume that this is a transcript error for "equilateral triangle," then what Mr. Goodwin of Solitude, Posey County, had discovered in his first statement was the equivalent of $\pi = 16/\sqrt{3} = 9.2376\ldots$, which probably represents the biggest overestimate of π in the history of mathematics.

ENGROSSED HOUSE BILL

No. 246

Read first time Jany. 15th 1897

Referred to Committee on

Canals - rep. and referred to Com on Education Jany. 19th 1897

Reported back Feby. 2d 1897

Read second time Feby. 5th 1897

Ordered engrossed Feby. 5th 1897

Read third time Feby. 5th 1897

Passed February 5th 1897

Ayes — 67 — Noes — 0 —

Introduced by Record,

IN THE SENATE.

Read first time and referred to Com. on Temperance. 2/11/97 Reported favorable 2/12/97 Read second time and indefinitely postponed 2/12/97

Facsimile of Bill No. 246, Indiana State Legislature, 1897. Kindly made available by the Indiana State Library.

However, Sections 1 and 2 contain more hair-raising statements which not only contradict elementary geometry, but also appear to contradict each other. Section 2 of the bill concludes

> *By taking the quadrant of the circle's circumference for the linear unit, we fulfill the requirements of both quadrature and rectification of the circle's circumference. Furthermore, it has revealed the ratio of the chord and arc of ninety degrees, which is as seven to eight, and also the ratio of the diagonal and one side of a square which is as ten to seven, disclosing the fourth important fact, that the ratio of the diameter and circumference is as five-fourths to four; and because of these facts and the further fact that the rule in present use fails to work both ways mathematically, it should be discarded as wholly wanting and misleading in practical applications.*

to the square on a line equal to the quadrant of the circumference, as the area of an equilateral rectangle is to the square on one side. The diameter employed as the linear unit according to the present rule in computing the circles area is entirely wrong, as it represents the circle's area one and one-fifth times the area of a square whose perimeter is equal to the circumference of the circle. This is because one-fifth of the diameter fails to be represented four times in the circle's circumference. For example: if we multiply the perimeter of a

From a page of the original copy of Bill No. 246, Indiana State Legislature, 1897. Kindly made available by the Indiana State Library.

And Section 3:

> *In further proof of the value of the author's proposed contribution to education, and offered as a gift to the State of Indiana, is the fact of his solutions of the trisection of the angle, duplication of the cube and quadrature of the circle having been already accepted as contributions to science by the American Mathematical Monthly, the leading exponent of mathematical thought in this country. And be it remembered that these noted problems had been long since given up by scientific bodies as unsolvable mysteries and above man's abilities to comprehend.*

The bill was, perhaps symbolically, referred to the House Committee on Swamp Lands, which passed it on to the Committee of Education, and the latter reported it back to the House "with recommendation that said bill do pass." On February 5, 1897, the House passed the learned treatise unanimously (67 to 0).

Five days later the bill went to the Senate, where it was referred, for unknown reasons, to the Committee on Temperance. The Committee on Temperance, too, reported it back to the Senate with the recommendation that it pass the bill, and it passed the first reading without comment.

What would have happened to the bill if events had run their normal course is anybody's guess. But it so happened that Professor C.A. Waldo, a member of the mathematics department of Purdue University, was visiting the State Capitol to make sure the Academy appropriation was cared for, and as he later reported,[90] was greatly surprised to find the House in the midst of a debate on a piece of mathematical legislation; an ex-teacher from eastern Indiana was saying: "The case is perfectly simple. If we pass this bill which establishes a new and correct value of π, the author offers our state without cost the use of this discovery and its free publication in our school textbooks, while everyone else must pay him a royalty."

Professor Waldo, horrified that the bill had passed the House, then coached the senators, and on its second reading, February 12, 1897, the Senate voted to postpone the further consideration of this bill indefinitely; and it has not been on the agenda since.

ANOTHER modern American circle-squarer was one John A. Parker, who published a book in New York in 1874[91] called *The*

Quadrature of the Circle. Containing Demonstrations of the Errors of Geometers in Finding Approximations in Use. The book contains 218 pages on the quadrature of the circle, to which is added more of the author's wisdom on polar magnetism and other matters.

> The great and fatal error of geometers (writes Parker) is seen in their declaration that the circle and the square are "incommensurable," that there is no co-relation between circumference and diameter, and our first proposition shall be to dissipate this fatal error; therefore
>
> PROPOSITION FIRST
>
> The circumference of any circle being given, if that circumference be brought into the form of a square, the area of that square is equal to the area of another circle, the circumscribed square of which is equal in area to the area of the circle whose circumference is first given.

Mr. Parker must have found the rejection of his theories very frustrating, because the above proposition (as is easily checked with paper and pencil) is perfectly correct; it is also perfectly trivial. Like most other modern circle-squarers, he did not even understand the problem; how, for example, was he going to construct a square whose perimeter is equal to the circumference of the circle? He substitutes numbers in his proposition, and thinking that the result constitutes a proof, goes on to say

> For if the circumference and diameter of a circle be really *incommensurable*, as geometers have affirmed, then no circle and square *can* be exactly equal to one another. But when it has been demonstrated as has here been done, that a circle and a square may be exactly equal one to the other, then it is demonstrated that the two are *not incommensurable*; and with this demonstration the whole theory of mathematicians . . . is proved to be fallacious.

Elsewhere Parker "demonstrates" a theorem according to which $\pi = 20612 : 6561$ exactly. "And that proposition being proved," he exclaims, "all the serial and algebraic formula in the world, or even geometrical demonstration, if it be subjected to any error whatever, cannot overthrow the ratio of circumference to diameter which I have established!" His triumph makes him generous, and he praises Metius for using the ratio 355/113 "more than a century past" (Metius lived in the 16th century), which is "the nearest approximation to the truth ever made in whole numbers."

The book contains more of Parker's wisdom, such as the intriguing theorem that "The circumference of a circle is the line *outside* of the circle thoroughly inclosing it." This theorem proves to be the undoing of all other geometers who use a line "coinciding with the utmost limit of the area of the circle," instead of "inclosing" it as Parker does, so that "by this difference, with their approximation, geometers make an error in the sixth decimal place."

More interesting than Parker's mathematical garbage, perhaps, is the psychology of his attitude to professional mathematicians. Throughout the ages, men with little education, small intellect, or both, have sometimes felt an envious dislike of their intellectual superiors, branding them as "effete snobs," "eggheads," and the like. Parker expresses this attitude as follows:

> I have found the Professors as a body, though learned in received theories, to be among the *least competent* to decide on any newly discovered principle. Their interest, education, pride, prejudice, self-love and vanity, all rise in resistance to anything which conflicts with their tenets, or which outruns the limits of their own reasoning. So little do they look beyond the principles inculcated by education, and so tenaciously do they hold on to these, that when driven from one principle, they fall back to another, and when beaten from all, they return again to the first, and maintain themselves by dogged assertion, or charging their assailants with ignorance and a lack of science.
>
> In consequence of this character of professors, the practical men of the age are at least a century in advance of the schools in all useful scientific knowledge.

JOHN A. Parker was, however, a mild case of circle-squaromania compared to one Carl Theodore Heisel, who described himself as Citizen of Cleveland, Ohio, and 33° Mason, and who in 1931 published a book whose title pages are reproduced on the next two pages. Not content with squaring the circle, he also rejects decimal fractions as inexact (whereas ratios of integers are exact and scientific), "disproves the world renowned Pythagorean problem accepted as absolute by geometricians for the last twenty-four centuries," and extracts roots of negative numbers thus: $\sqrt{-a} = \sqrt{a-}$; $\sqrt{a-}^2 = -a$. As for squaring the circle, his grand discovery amounts to setting $\pi = 256/81$, the very same value used by Ahmes the Scribe in Egypt some 4,000 years before Heisel. Substituting this value for calculations of areas and circumferences of circles with diameters 1, 2, 3,. . . up to 9, he obtains numbers showing consistency of circumference and area, "thereby furnishing incontrovertible proof of the exact truth" of his ratio, never noticing that he would have obtained the same consistency had he set π equal to the birthdate of his grandmother.

From this it is probably clear that "this great American discovery, which cannot fail to be immortal, because absolutely irrefutable!" is not very interesting; but Heisel's book throws considerable light on

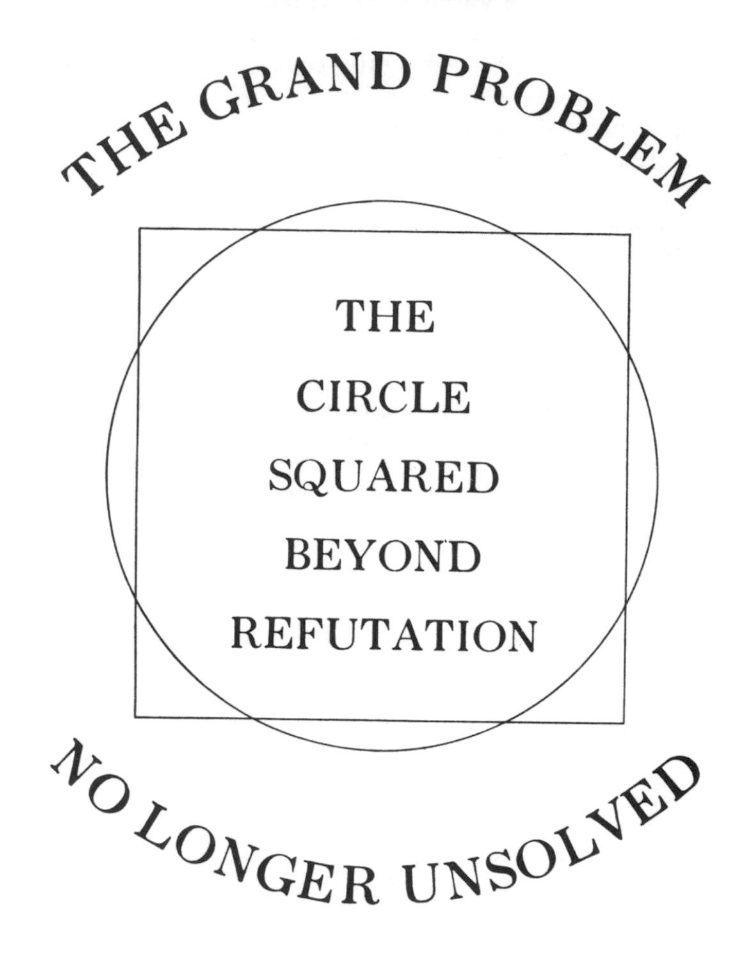

Title pages of Heisel's book.

the mentality of the genius whose great revelations are not appreciated by a prejudiced and ignorant world.

If his circle ratio is the same as that used in ancient Egypt, then his self-appraisal is reminiscent of Esarhaddon, King of Assyria in the 7th century B.C., who began an inscription carved in stone with "I am

Mathematical
and
Geometrical Demonstrations

By
Carl Theodore Heisel

Disproving Numerous Theorems, Problems, Postulates, Corolleries, Axioms, and Propositions, with Ratios, Laws and Rules Hitherto Unknown in Mathematical and Geometrical Science, Naturally Growing Out of the

Extraordinary and Significant Discoveries of a

LACKING LINK

by Carl Theodore Faber
Who was a Citizen of Brooklyn, N. Y., U. S. A.

IN THE DEMONSTRATION OF THE WORLD RENOWNED

PYTHAGOREAN PROBLEM

Utterly Disproving Its Absolute Truth, Although Demonstrated as such for Twenty-Four Centuries; and by This Discovery Establishing the Fact of the Existence of Perfect Harmony Between Arithmetic and Geometry as a Law of Nature, and Calculating to Settle Forever the Famous Dispute Between

The Two Great Philosophical Schools

By CARL THEODORE HEISEL
A Citizen of Cleveland, Ohio, U. S. A.

FIRST EDITION
1931

Title pages of Heisel's book.

powerful, I am omnipotent, I am a hero, I am gigantic, I am colossal!" Heisel's biography, facing a full-page photograph captioned

CARL THEODORE HEISEL, 33°

and written by himself, informs us that

> He is a man of stalwart figure, strong personality, and splendid physical and mental vigor. He possesses excellent health and has an abundance of energy. He has been a successful business man, an accountant, a student, an investigator and a practical philosopher . . . He has read exhaustively and has traveled extensively.
>
> Illustrious Brother Carl Theodore Heisel is a 33° Mason. He has been a serious student, a thorough investigator, a capable organizer, and a tireless worker in the various degrees of that order . . .
>
> He is not publishing this book with any idea of profit from the sale of such book. He is addressing it more especially to the practical student, who with common sense prefers to do his own thinking, and not blindly accept the opinions and teachings of professors of mathematics and geometry, who fear being ridiculed and ostracized by the profession generally, and losing their positions as professors or teachers, if they acknowledge any truth not found in the books as taught today . . .
>
> He believes that [his deceased colleague] Mr. Theodore Faber has solved the great problem, and that his ratios are absolutely true, accurate, incontrovertible, and to be the greatest mathematical discovery in the history of the world.

But Heisel outdoes himself when he accuses a professor of the University of Cambridge of having stolen his great mathematical discovery after Heisel had communicated it to the Royal Society in London. "If the American public should ever wake up to the importance of the discovery," writes Heisel, "and to the national honor reflected by it, this singular coincidence may become a question of international investigation."

Heisel was evidently a man of considerable means, for he had several thousand copies of the book printed and distributed free to libraries, colleges and scientists, and there is probably a copy of this curiosity in most American college libraries that he considered important in 1931.

18

The Computer Age

> ... the foulest thing I have ever read. That these "scientists" should toy with problems of such universal and profound significance shows only that our educational system is an abominable failure, turning out unnatural, immoral and monstrous specimens of humanity. Are these creatures now to change our lives in the name of science?
>
> From a reader's letter to TIME

THE story of π in the computer age of the 20th century is reminiscent of that of the digit hunters in the 18th and 19th centuries. The main difference is that where the digit hunters of the 18th and 19th centuries topped the standing records by tens and hundreds of decimal places, the computers and their programmers topped the standing records by thousands, and then by hundreds of thousands of digits. By 1967, the value of π was known to 500,000 decimal places. And, of course, where the digit hunters had drudged for months and years to find hundreds of decimal places, the computer that churned out half a million digits needed only 26 hours and 40 minutes (plus 1 hour and 30 minutes to convert the final result from binary to decimal notation). The similarity between the *idiots savants* of the 18th and 19th centuries and the imbecility of the 20th century computer has already been pointed out (p. 105). Yet as we come to the end of the story, this similarity vanishes; for we are living at a time when some computers (more accurately, their programs) have become remarkably intelligent.

Before we take a look at these lately arrived intelligent computers, we return to their older, moronic brother, who is not capable of anything but slavishly following comparatively simple commands of its programmer, albeit with enormous speed and the use of its vast memory. It was this kind of computer that was used to rattle off the decimal digits of π. The details of the story have been told by Wrench,[93] and only a brief survey will be given here.

The first computer calculation of π was apparently made in September 1949 on ENIAC (Electronic Numerical Integrator and Computer) at the Ballistic Research Labs; it calculated π to 2,037 places in 70 hours, a pitifully long time by today's standards. Like many other computer evaluations, this one was programmed in accordance with Machin's formula (14), p. 144, in the form

$$\pi = 16 \arctan(1/5) - 4 \arctan(1/239) \qquad (1)$$

In November 1954 and January 1955, NORC (Naval Ordnance Research Calculator) at Dahlgren, Virginia, was programmed to compute π to 3,089 significant places; the run took only 13 minutes.

This record was broken at the Ferranti Computer Centre, London, in March 1957, when a Pegasus computer computed 10,021 decimal places in 33 hours. The program was based on an arctangent formula similar to, but not identical with, the one used by Strassnitzky (p. 107). However, a subsequent check revealed that a machine error had occurred, so that "only" 7,480 decimal places were correct. The run was therefore repeated in March 1958, but the correction was not published.

Then, in July 1958, an IBM 704 at the Paris Data Processing Center was programmed according to a combination of Machin's formula and the Gregory series, corresponding to (15) on p. 144; it yielded 10,000 decimal places in 1 hour and 40 minutes.

A year later, in July 1959, the same program was used on an IBM 704 at the Commissariat à l'Energie Atomique in Paris, and 16,167 places were obtained in 4.3 hours.

Machin's formula was also the basis of a program run on an IBM 7090 at the London Data Centre in July 1961, which resulted in 20,000 decimal places and required only 39 minutes running time.

By this time the limit of the then available computer memories had almost been reached. Further substantial increases in the number of decimal places could have been obtained only by modifying the programs to use more machine time and therefore to run into unreasonable costs.

But in July 1961, Shanks and Wrench[94] increased the speed of the computation by a factor of about 20. (Daniel Shanks, incidentally, is not related to William Shanks, who calculated 707 places just 100 years ago, see p. 103). In part, this was due to a faster computer (an IBM 7090 at the IBM Data Processing Center, New York), but they also used several tricks in programming it; in particular, they abandoned Machin's formula in favor of the formula

$$\pi = 24 \arctan(1/8) + 8 \arctan(1/57) + 4 \arctan(1/239) \qquad (2)$$

which was found by Störmer in 1896. The run resulted in 100,265 decimal places, of which the first 100,000 were published[94] by photographically reproducing the print-out with 5,000 decimals per page. The first 10,000 places of the print-out are reproduced on the end sheets of this book. The time required for computing the first term in (2) was 2 hours and 7 minutes, for the second term 3 hours and 7 minutes, and for the third term 2 hours and 20 minutes. To this must be added 42 minutes for converting the final result from binary to decimal digits, so that the total time required was 8 hours and 43 minutes.

A computation of this kind involves billions of individual arithmetic operations, and if a single one of these is mistaken, the entire subsequent operation may yield an erroneous result. It is therefore necessary to check the result. For this, Shanks and Wrench used a special method which calculates π by a different formula (another arctangent formula, due to Gauss), but uses the partial results of the original run in such a way that the check takes less time than the original computation.

Subsequently, π was computed to 250,000 decimal places on an IBM 7030 at the Commissariat à l'Energie Atomique in Paris in February 1966, and a year later, in February 1967, a CDC 6600 was programmed by J. Gilloud and J. Filliatre, at the same institution, to yield 500,000 decimal places. The program was again based on Störmer's formula (2) and the Shanks-Wrench method for checking the digits; the running time was 28 hours and 10 minutes (of which 1 hour and 35 minutes were used for conversion), and an additional 16 hours and 35 minutes were needed for the check. These quarter- and half-million digit values of π were published in reports of the Commissariat à l'Energie Atomique in Paris.

This, as far as I know, is the present record. I may be mistaken, and even if I am not, this record will, no doubt, eventually be broken.

The driving force behind these computations seems to be, at least in part, the same as the one that drove Ludolf van Ceulen to find the first 20 decimal places in 1596. Yet these hundreds of thousands of digits are not quite as useless as the results of the earlier digit hunters. There are two reasons for this. The first, admittedly, is not very convincing. It concerns the statistical distribution of the digits, which is expected to be uniform, that is, the frequency with which the digits (0 to 9) appear in the result will tend to the same limit (1/10) as the number of decimal places increases beyond all bounds. An analysis of the first 16,000 decimal digits bears this out within the usual statistical tests,[93] but this does not constitute a rigorous proof for a finite number of digits, no matter how large; on the other hand, a rigorous theoretical proof (which has not yet been given) has no need of the actual arithmetical computation. And as for the generation of digits with equal probabilities, this can be done in much simpler ways.

The other reason for such computations is more convincing. Before it goes into operation, a computer, like any other machine, is tested whether it can do its job reliably. One such method is to let it churn out a few tens of thousands of decimal digits of π and to check the result against the known figures; if they agree, the computer has performed millions of arithmetical operations faultlessly. (There are, of course, other functions that must also be tested.)

ALL of the computations above were performed by computers with not an ounce of intelligence. The frustrations resulting from the computer's inability to insert a simple dot have been remarked on before (p. 105). However, it should be added that it takes only a relatively primitive program to make the computer supply the missing dot and to print, for example, the following comment:

LOOK BUDDY, I PUT IN A DOT FOR YOU IN LINE 123, BUT NEXT TIME DO IT YOURSELF, OK?

But this does not, of course, constitute intelligence. Every step to produce this result must be covered by the instructions that make a computer execute a program, and the above sentence must be, so to speak, put into the computer's mouth by the programmer. That is not the way one gives instructions to an intelligent being. If you ask your wife (or husband) to bring you a glass of water, you don't instruct her (or him) exactly what muscle to move at any given time. She will, without specific instructions, turn on the cold, not the hot, water, and she will use her own judgement in unexpected situations — if for some reason no glasses are available, she will overrule your instructions and bring you a cup, even though you asked for a glass. That does not take much intelligence, but it is a lot more than most contemporary computers have.

Will computers ever become intelligent?

They already have. Not the morons that bill your charge account or that compute the decimal places of π, but the amazing programs (it is the programs rather than the actual computer hardware) that have been growing in the last few years at Stanford, M.I.T., Johns Hopkins, and other laboratories.

Intelligence, says my dictionary,[96] is "the ability to adapt to new situations, and to learn from experience; the inherent ability to seize the essential factors of a complex matter."

Believe it or not, but there is nothing in that definition that a machine cannot be programmed to do. Programs have been written that learn from experience, adapt to new conditions, grasp the essentials of a complicated problem, and decide for themselves how to solve it; and all that (as yet in a few very restricted areas) with an intelligence that approaches that of the best humans in the field, and far surpasses the intelligence of most others. The stress here is on how *well* they can do this, not on how many varied problems they can manage, for the memory of a computer and the access to it cannot (yet) compete with the human brain. But the qualitative principle is there.

Take, for example, the program that plays checkers, as developed over the years since 1947 by Arthur Samuel.[97] This program will learn from experience (Samuel improved it by making several computers play checkers furiously against each other for prolonged periods). It can also learn from other players' experience, and it will "study" other people's games and moves recommended by champions. In a given position, it will not slavishly go through all the possible moves and their consequences (there are too many), but it will use certain criteria to evaluate its own position and to determine the best strategy, and it will then make its own decision as to the next move. The results: Although the program could not beat the world checkers champion, it did beat the champion of Connecticut, it would probably beat you, it would certainly beat me, and — an extremely significant fact — *it beats its own programmer.*

There are many other examples. There are programs to play other games (including chess) intelligently, programs that will prove theorems (one such program proved, in its initial version, 38 out of the first 52 theorems in Newton's *Principia*), programs that verify mathematical proofs and expose fallacies, programs that solve general problems to attain given goals, and many others, including one with great potentialities: a program to write programs.[97] A computer that is particularly dramatic, though perhaps less sophisticated than others in this class, is The Beast, a battery-operated cylinder on wheels built by scientists at the Applied Physics Lab of Johns Hopkins University. It has its own computer logic and steering, and it is furnished with tactile, optical and sonar sensors. The Beast was often let loose to roam the halls and offices of the Applied Physics Lab, which it would do without bumping into walls or falling downstairs (it would turn round on sensing a step), and when its batteries were low, The Beast would optically find an outlet in some office, plug itself in, and depart again when it had "eaten," no doubt often leaving behind a new secretary frozen in horrified incredulity.[97]

But let us return to the checkers program that can beat its own programmer. A long time ago, even when he constructed his first bow and arrow, man used his intelligence to design machines that surpassed him in speed, force, and many other qualities. Arthur Samuel's program might be taken as an historic landmark: Somewhere near that point, man first used his intelligence to design a machine that surpassed him in intelligence. We are now only at the birth of such a machine, but eventually the intelligent computer might be to the moronic computer as the spacecraft is to the bow and arrow. There are already programs to write programs, and programs to balance assembly lines. It is therefore entirely within the realm of possibility that such a machine will eventually have the ability to reproduce itself.

"Destroy it!" is what the pious, respectable and community-minded ladies will scream when word gets out about the new computer.

Their screams have been heard before.

"Destroy it!" is what Julius Caesar screamed as his hordes put the torch to the Library of Alexandria.

"Destroy it!" is what the Grand Inquisitor screamed when he read Galileo's *Dialogues*.

"Destroy it!" is what the Luddites screamed in 18th-century England when they smashed the machinery that was supposedly responsible for their misery in the Industrial Revolution.

"Destroy it!" is what the Soviet censor screams when he sees a copy of Orwell's *1984*.

"Destroy it!" is what the Fascists of the Left screamed when they bombed or smashed computing centers in Minnesota or Montreal.

It has again become fashionable to blame science and technology for the ills of society. I have some sympathies for the Luddites who were uneducated, miserable, and desperate. I have none for the college-educated illiterates who drivel about "too much science and technology" because they want to conserve their life style by denying it to everybody else.

THREE centuries ago, Gottfried Wilhelm Leibniz, co-inventor of the calculus and co-discoverer of the first infinite series for π, dreamt of the day when courts would be abolished, because disputes would be settled mathematically by solving impartial equations that would show who was right and who was wrong. The intelligent computer that is now being born makes that dream somewhat less fantastic. Perhaps the *n*th generation of intelligent computers will make a better job of keeping peace among men and nations than men have ever been able to.

And with that thought our story of π is coming to an end. It is a story as varied as the brilliance of Archimedes of Syracuse and the ignorance of Heisel of Cleveland.

The history of π is only a small part of the history of mathematics, which itself is but a mirror of the history of man. That history is full of patterns and tendencies whose frequency and similarity is too striking to be dismissed as accidental. Like the laws of quantum mechanics, and in the final analysis, of all nature, the laws of history are evidently statistical in character.

But what those laws are, nobody knows. Only a few scraps are evident. And one of these is that the Heisels of Cleveland are more numerous than the Archimedēs of Syracuse.

NOTES

1. Dantzig.
2. Kolman
2. Kolman.
3. *Sagrada Biblia,* Editorial Catolica, Madrid, 1955.
4. From Sato Moshun's *Tengen Shinan* (1698), see Smith and Mikami, p.131.
5. Neugebauer.
6. Neugebauer, pp.58-61.
7. Midonick.
8. Midonick.
9. Rudio, p.18
10. Rudio, p.19
11. See bibliography.
12. Needham, p.29. Reproduced by kind permission of Cambridge University Press.
13. Needham, p. 43. Reproduced by kind permission of Cambridge University Press.
14. Hogben (1937).
15. Brandon.
16. Collier.
17. Butkevich *et al.* The quotation has been retranslated from Russian and may not be quite accurate.
18. De Camp, who quotes Bishop Landa.
19. Coolidge (1949), pp. 46-47.
20. Oxford, 1931. Reprinted by E.P. Dutton & Co., New York, 1967.
21. According to some, the Museum and Library were founded by Ptolemy I, but acquired its definite form under Ptolemy II. Others still say that the Museum was founded by Arsinoe.
22. De Camp.
23. It could be proved by a *reductio ad absurdum*, but so could many other things that do not follow from Euclid's axioms, e.g., the statement that "cause precedes effect."
24. De Camp
25. Loeb Classical Library, 10 vols., 1938-63.
26. See Cantor, vol. I, Chapters 25-27, for a very poor record compared with that of their contemporaries.
27. Rudio.
28. Barnes, vol. 1, p. 255.
29. Geddie.
30. *Saint Joan*
31. Lodge, p. 61.
32. From *On Floating Bodies*, translated by Sir Thomas Heath, see bibliography under Heath (1897 and 1912). Reproduced by kind permission of Cambridge University Press.
33. From *Aristotle's Physics*, translated by H.G. Apostle, © 1969 by Indiana University Press, reproduced by permission of the publishers.

34. Heath (1897 and 1912).
35. Tropfke, p. 210.
36. A hand was about 4 inches, a cubit about 21.8 inches (J.P. Boyd, *Bible Dictionary*, Ottenheimer Publishers, Owing Mills, Md., 1958).
37. Hogben (1937).
38. It is even doubtful whether Pythagoras knew Pythagoras' Theorem, and if so, whether he could prove it. See Tropfke, pp. 137-138.
39. Depman. In fairness I must add that I have not found a reference to this event anywhere else, and that Soviet books are unreliable where competitive religions are concerned.
40. Hogben (1937).
41. De Camp.
42. Lodge.
43. Neugebauer.
44. Hogben (1937) writes that there was also a stream of Jewish physicians who brought the science of Algebra to Europe, pointing to the term "Physician and Algebraist" used in Spain at the time, and I used this interpretation in previous editions of this book. Not so, writes Dr. R.B. Lees, professor of linguistics at Tel-Aviv University, to whom I am greatly indebted for this and other comment. Arabic *el jabr* meant "the bonesetting" and only later came to be used in the sense of reuniting the parts of an equation.
45. Hogben (1937); but Boyer (1968) says it is not clear where Adelard came into contact with Muslim learning.
46. Boyer (1968).
47. Tropfke.
48. Boyer (1968)
49. Cantor, vol. 2, pp. 199-201.
50. Incorrectly attributed to Galileo (see Lodge, p. 131); may not be true of Bruno, either.
51. De Camp.
52. Ore.
53. *Zeitschrift f. Mathematik und Physik*, vol. 36, *Historisch-literarische Abteilung*, pp. 139-140 (1891).
54. Ore.
55. See Chapter 18.
56. Schubert. The distance to Sirius has been brought up to date.
57. Tropfke, p. 217.
58. Newman, vol. 1, p. 466.
59. Ball.
60. Courtesy of Miss Angela Dunn, Director, Problematical Recreations, Litton Industries, Beverly Hills, California.
61. See bibliography.
62. Needham, Fig. 80, p. 135. Reproduced by courtesy of Cambridge University Press.
63. Freyman.
64. Boyer (1968), p. 401.
65. Smith and Mikami, p. 87.
66. Smith and Mikami, p. 130.
67. This follows Hogben's suggestion, see Hogben (1937), p. 258-262.
68. Struik (1969), pp. 244ff.

69. See, for example, C.D. Olds, *Continued Fractions*, Random House, N.Y., 19€3.
70. Tropfke, p. 224.
71. Lodge.
72. Andrade.
73. Newton, *Treatise on the Method of Fluctions and Infinite Series*, London, 1737, pp. 129-131. See Whiteside, vol. I, pp. 100-101.
74. Smith (1929), p. 346.
75. *Leonard Euler, Perepiska* (Leonhard Euler, Correspondence), Nauka, Leningrad, 1967.
76. Freyman.
77. Tremble, enemies of France,
 Kings drunk with blood and conceit!
 The sovereign people advances,
 Tyrants, descend into the grave!
78. Hammersley and Hanscomb.
79. Beckmann, P., *Elements of Applied Probability Theory*, Harcourt, Brace & World, New York, 1968; pp. 37-38.
80. On the Librascope of the Electr. Engrg. Dept., Univ. of Colorado.
81. This line of reasoning works with sufficient, not necessary conditions. Actually, the final equation will be quadratic, since it can be shown that each new step results in a quadratic equation with coefficients that are either rational or square roots. For details, see Hobson, pp. 47-51.
82. From *A Concise History of Mathematics* by Dirk J. Struik, Dover Publications, Inc., New York. Reproduced by permission of the Publisher.
83. Coxeter, p. 53.
84. Tropfke, p. 217
85. Lindemann, F: "Über die Zahl π," *Mathematische Annalen*, vol. 20, pp. 212-225 (1882); also "Über die Ludolphsche Zahl," *Berichte der Berliner Akademie*, vol. 2, pp. 679-682 (1882).
86. Weierstrass, K.W.: "Zu Lindemanns Abhandlung 'Über die Ludolphsche Zahl'," *Berichte der Berliner Akademie*, vol. 5,pp.1067-1085 (1885).
87. Hobson, pp. 53-57.
88. Schubert, p. 143.
89. The jacket of the handwritten copy is thus entitled, and with the same capitalization.
90. W.E. Eddington, "House Bill No. 246, Indiana State Legislature, 1897," *Proc. of the Indiana Academy of Sciences*, vol. 45, pp. 206-210 (1935).
91. See bibliography under *Parker*.
92. Before the advent of the computer, that is. The natural logarithm and square root of 2 are now known to 3,683 and 1,000,082 decimals, respectively. See note 95.
93. Wrench (1960).
94. Shanks and Wrench (1962).
95. This paragraph (and also note 92) is based on a private communication by Dr. J.W. Wrench, Jr., which is gratefully acknowledged.
96. *Funk and Wagnall's Standard College Dictionary*, Harcourt, Brace & World, New York, 1963.
97. J.R. Slagle, *Artificial Intelligence: The Heuristic Programming Approach*, McGraw-Hill, New York, 1971.

BIBLIOGRAPHY

ANDRADE, E.N. da C., *Sir Isaac Newton*, Doubleday, New York, 1958.

BALL, W.W.R., *A Short Account of the History of Mathematics*, Macmillan, London, 1888. Reprinted by Dover Publications, 1960.

BARNES, H.E., *An Intellectual Cultural History of the World*. In 3 vols., 3rd ed., Dover Publications, New York, 1965.

BELL, E.T., *Men of Mathematics*, Simon and Schuster, New York, 1937.

— — —, *The Development of Mathematics*, McGraw-Hill, New York, 1940.

BOYER, C.B., *The History of the Calculus and Its Conceptual Development*, 2nd ed., Hafner, New York, 1949; reprinted by Dover Publications, 1959.

— — —, *A History of Mathematics*, Wiley, New York, 1968.

BRANDON, W., *The American Heritage Book of Indians*, American Heritage Publ. Co., New York, 1961; reprinted by Dell Publ. Co., New York, 1964.

BUTKEVICH, A.V., GANSHIN, V.N., KHRENOV, L.S., *Vremya i kalendar'* (Time and the Calendar), Vysshaya shkola, Moscow, 1961.

CANTOR, M., *Vorlesungen über Geschichte der Mathematik*, Teubner, Leipzig; vol. 1, 1894, vol.2, 1900; vol. 3, 1901; vol. 4, 1908.

COLLIER, J., *Indians of the Americas*, New American Library (Mentor Books), New York, 1948.

COOLIDGE, J.L., *The Mathematics of Great Amateurs*, Clarendon Press, Oxford, 1949; reprinted by Dover Publications, 1963.

— — —, *A History of the Conic Sections and Quadric Surfaces*, University Press, Oxford, 1945; reprinted by Dover Publications, 1968.

COXETER, H.S.M., *Introduction to Geometry*, Wiley, New York, 1969.

DANTZIG, T., *Number: The Language of Science*, London, 1940; Macmillan, New York, 1943.

DE CAMP, L.S., *The Ancient Engineers*, Doubleday, Garden City, N.J., 1963.

DEPMAN, I.Y., *Rasskazy o matematike* (Tales about Mathematics), Gosdetizdat, Leningrad, 1954.

EDINGTON, E., *House Bill No. 246, Indiana State Legislature, 1897*. Proceedings of the Indiana Academy of Sciences, vol. 45, pp. 206-210 (1935).

FREYMAN, L.S., *Tvortsy vysshey matematiki* (Creators of Higher Mathematics), Nauka, Moscow, 1968.

GEDDIE, W., and GEDDIE, L. (Ed.), *Chambers Biographical Dictionary*, 2nd ed., Chambers, London, 1938.

GREENBLATT, M.H., *The "legal" value of π and some related mathematical anomalies*, American Scientist, vol. 53, pp. 427A-432A, Dec. 1965.

HAMMERSLEY, J.M., and HANDSCOMB, D.C., *Monte Carlo Methods*, Wiley, New York, 1964.

HEATH, T.L., *The Works of Archimedes*, Cambridge University Press, Cambridge, 1897 and 1912. Reprinted by Dover Publications, 1953.

— — —, *The Thirteen Books of Euclid's Elements*, 3 vols., Cambridge University Press, Cambridge, 1908. Reprinted by Dover Publications, 1956.

— — —, *A Manual of Greek Mathematics*, Oxford University Press, Oxford, 1931. Reprinted by Dover Publications, New York, 1963.

HEISEL, C.T., FABER, C.T., *Behold! The Grand Problem. The Circle Squared Beyond Refutation. No Longer Unsolved*, (Printed by) S.J. Monck, Cleveland, Ohio, 1931.

HOBSON, E.W., *Squaring the Circle*. Oxford University Press, 1913. Reprinted by Chelsea Publ. Co., New York (no date given).

HOGBEN, L., *Mathematics for the Million*, W.W. Norton & Co., New York, 1937; reprinted by Pocket Books, New York, 1965.

— — —, *Mathematics in the making*, Rathbone Books, London, 1960.

KOLMAN, E., *Istoriya matematiki v drevnosti* (History of Mathematics in Antiquity), Fizmatgizdat, Moscow, 1961.

LIETZMANN, W., *Methodik des mathematischen Unterrichtes*, vol. 2, Quelle & Meyer, Leipzig, 1923.

— — —, *Altes und Neues vom Kreis*, Teubner, Leipzig, 1935.

LINDEMANN, F., *Über die Ludolphsche Zahl*, Berichte der Berliner Akademie, vol. 2, pp. 679-682 (1882).

— — —, *Über die Zahl* π, Mathematische Annalen, vol. 20, pp. 221-225 (1882).

LODGE, Sir Oliver, *Pioneers of Science*, Macmillan & Co., London, 1893. Reprinted by Dover Publications, New York, 1960.

MAISTROV, L.E., *Teoriya veroyatnostey — istoricheski ocherk* (Probability Theory — An Historical Outline), Nauka, Moscow, 1967.

MIDONICK, H.O., (Ed.), *A Treasury of Mathematics*, Philosophical Library, New York, 1965.

MORLEY, S.G., *The Ancient Maya*, Stanford University Press, 1947.

NEEDHAM, J., *Science and Civilization in China*, vol. III, Cambridge University Press, Cambridge, 1959.

NEUGEBAUER, O., *The Exact Sciences in Antiquity*, 2nd ed., Brown University Press, 1957. Reprinted by Dover Publ., New York, 1969.

NEWMAN, J.R., *The World of Mathematics*, 4 vols., Simon & Schuster, N.Y., 1969.

ORE, O., *Cardano the Gambling Scholar*, Princeton University Press, Princeton, N.J., 1953. Reprinted by Dover Publications, New York, 1965.

PARKER, J.A., *Quadrature of the Circle*, John Wiley & Son, New York, 1874.

RAPPORT, S., WRIGHT, H., *Mathematics*, New York University Press, New York, 1963. Reprinted by Washington Square Press, New York, 1964.

RUDIO, F., *Archimedes, Huygens, Lambert, Legendre. Vier Abhandlungen über die Kreismessung*, Leipzig, 1892.

SCHUBERT, H., *The Squaring of the Circle*. Contained in *Mathematical Essays and Recreations*, Translation from German, Open Court Publishing Company, Chicago, 1899.

SIMON, M., *Uber die Entwicklung der Elementargeometrie im XIX. Jahrhundert*, Leipzig, 1906.

SHANKS, D., WRENCH, J.W., Jr., *Calculation of* π *to 100,000 decimals*, Mathematics of Computation, vol. 16, pp. 76-99, January 1962.

SMITH, D.E., *A Source Book in Mathematics*, 1929. Reprinted by Dover Publ., New York, 1959.

SMITH, D.E., and MIKAMI, Y., *A History of Japanese Mathematics*, Open Court Publishing Co., Chicago, 1914.

STRUIK, D.J., *A Concise History of Mathematics*, 3rd rev. ed., Dover Publications, New York, 1967.

— — —, *A Source Book in Mathematics, 1200-1800*. Harvard University Press, Cambridge, Mass., 1969

TROPFKE, J., *Geschichte der Elementarmathematik. Vierter Band: Ebene Geometrie*. Vereinigung wissenschaftlicher Verleger, Berlin-Leipzig, 1923.

TURNBULL, H.W., *The Great Mathematicians*, Simon & Schuster, New York, 1962.

WEIERSTRASS, K.W., *Zu Lindemanns Abhandlung "Uber die Ludolphsche Zahl,"* Berichte der Berliner Akademie, vol. 5, pp. 1067-1085 (1885).

WHITESIDE, D.T. (Ed.), *The Mathematical Works of Isaac Newton*, Johnson Reprint Corporation, New York, vol.1, 1964; vol. 2, 1967.

WILDER, R.L., *evolution of Mathematical Concepts*, Wiley, New York, 1968.

WILLIAMS, T.I., (Ed.), *A Biographical Dictionary of Scientists*, Wiley-Interscience, New York, 1969.

WRENCH, J.W., Jr., *The Evolution of Extended Decimal Approximations to* π, The Mathematics Teacher, vol. 53, pp. 644-650, December 1960.

ZVORYKIN, A.A. (Ed.), *Biograficheski slovar' deyateley estestvoznaniya i tekhniki* (Biographical Dictionary of Scientists and Engineers), Izd. Bolshoy Sovetskoy Entsiklopedii, Moscow, vol 1, 1958; vol. 2, 1959.

CHRONOLOGICAL TABLE

ca. 2000 B.C.	Babylonians use $\pi = 3\ 1/8$
ca. 2000 B.C.	Egyptians use $\pi = (16/9)^2 = 3.1605$
12th century B.C.	Chinese use $\pi = 3$
ca. 550 B.C.	I Kings vii, 23 implies $\pi = 3$
ca. 440 B.C.	Hippocrates of Chios squares the lune
ca. 434 B.C.	Anaxagoras attempts to square the circle
ca. 430 B.C.	Antiphon enunciates the principle of exhaustion
ca. 420 B.C.	Hippias discovers the quadratrix
ca. 335 B.C.	Dinostratos uses the quadratrix to square the circle
3rd century B.C.	Archimedes establishes $3\ 10/71 < \pi < 3\ 1/7$ and $\pi \approx 211875 : 67441 = 3.14163$
	Archimedes uses the Archimedean Spiral to rectify the circle
ca. 225 B.C.	Appolonius improves the Archimedean value, unknown to what extent
2nd century A.D.	Ptolemy uses $\pi = 377/120 = 3.14166...$
3rd century A.D.	Chung Hing uses $\pi = \sqrt{10} = 3.16...$
	Wang Fau uses $\pi = 142/45 = 3.1555...$
263 A.D.	Liu Hui uses $\pi = 157/50 = 3.14$
5th century	Tsu Chung-Chi establishes $3.1415926 < \pi < 3.1415927$
ca. 500	Aryabhatta uses $\pi = 62832/2000 = 3.1416$
6th century	Brahmagupta uses $\pi = \sqrt{10} = 3.16...$
1220	Leonardo of Pisa (Fibonacci) finds $\pi \approx 3.141818$
before 1436	Al-Kashi of Samarkand calculates π to 14 places
1450	Cusanus finds approximation for length of arc
1573	Valentinus Otho finds $\pi \approx 355/113 = 3.1415929$
1583	Simon Duchesne finds $\pi = (39/22)^2 = 3.14256...$
1593	François Viète finds π as an infinite irrational product
1593	Adriaen van Roomen finds π to 15 decimal places
1596	Ludolph van Ceulen calculates π to 32 places, later to 35 decimal places
1621	Snellius refines the Archimedean method
1654	Huygens proves the validity of Snellius' refinement
1655	Wallis finds an infinite rational product for π; Brouncker converts it to a continued fraction

1665–1666	Newton discovers the calculus and calculates π to at least 16 decimal places; not published until 1737 (posthumously)
1671	Gregory discovers the arctangent series
1674	Leibniz discovers the arctangent series for π
1705	Sharp calculates π to 72 decimal places
1706	Machin calculates π to 100 places
1706	Jones uses the symbol π for the circle ratio
1719	De Lagny calculates π to 127 places
1748	Euler publishes the *Introductio in analysin infinitorum*, containing Euler's Theorem and many series for π and π^2
1755	Euler derives a very rapidly converging arctangent series
1766	Lambert proves the irrationality of π
1775	Euler suggests that π is transcendental
1794	Legendre proves the irrationality of π and π^2
1794	Vega calculates π to 140 decimal places
1840	Liouville proves the existence of transcendental numbers
1844	Strassnitzky and Dase calculate π to 200 places
1855	Richter calculates π to 500 decimal places
1873	Hermite proves the transcendence of e
1873-74	Shanks calculates π to 707 decimal places
1882	Lindemann proves the transcendence of π
1945	Ferguson finds Shanks' calculation erroneous from the 527th place onward
1946	Ferguson publishes 620 decimal places
1947	Ferguson calculates 808 places using a desk calculator
1949	ENIAC is programmed to compute 2,037 decimals
1954-1955	NORC is programmed to compute 3,089 decimals
1957	Pegasus computer (London) computes 7,480 places
1959	IBM 704 (Paris) computes 16,167 decimal places
1961	Shanks and Wrench improve computer program for π, use IBM 7090 (New York) to compute 100,000 decimal places
1966	IBM 7030 (Paris) computes 250,000 decimal places
1967	CDC 6600 (Paris) computes 500,000 decimal places

Index

PI = 3.+

1415926535 8979323846 2643383279 5028841971 6939937510 5820974944 5923078164 0628620899 8628034825 3421170679
8214808651 3282306647 0938446095 5058223172 5359408128 4811174502 8410270193 8521105559 6446229489 5493038196
4428810975 6659334461 2847564823 3786783165 2712019091 4564856692 3460348610 4543266482 1339360726 0249141273
7245870066 0631558817 4881520920 9628292540 9171536436 7892590360 0113305305 4882046652 1384146951 9415116094
3305727036 5759591953 0921861173 8193261179 3105118548 0744623799 6274956735 1885752724 8912279381 8301194912
9833673362 4406566430 8602139494 6395224737 1907021798 6094370277 0539217176 2931767523 8467481846 7669405132
0005681271 4526356082 7785771342 7577896091 7363717872 1468440901 2249534301 4654958537 1050792279 6892589235
4201995611 2129021960 8640344181 5981362977 4771309960 5187072113 4999999837 2978049951 0597317328 1609631859
5024459455 3469083026 4252230825 3344685035 2619311881 7101000313 7838752886 5875332083 8142061717 7669147303
5982534904 2875546873 1159562863 8823537875 9375195778 1857780532 1712268066 1300192787 6611195909 2164201989

3809525720 1065485863 2788659361 5338182796 8230301952 0353018529 6899577362 2599413891 2497217752 8347913151
5574857242 4541506959 5082953311 6861727855 8890750983 8175463746 4939319255 0604009277 0167113900 9848824012
8583616035 6370766010 4710181942 9555961989 4676783744 9448255379 7747268471 0404753464 6208046684 2590694912
9331367702 8989152104 7521620569 6602405803 8150193511 2533824300 3558764024 7496473263 9141992726 0426992279
6782354781 6360093417 2164121992 4586315030 2861829745 5570674983 8505494588 5869269956 9092721079 7509302955
3211653449 8720275596 0236480665 4991198818 3479775356 6369807426 5425278625 5181841757 4672890977 7727938000
8164706001 6145249192 1732172147 7235014144 1973568548 1613611573 5255213347 5741849468 4385233239 0739414333
4547762416 8625189835 6948556209 9219222184 2725502542 5688767179 0494601653 4668049886 2723279178 6085784383
8279679766 8145410095 3883786360 9506800642 2512520511 7392984896 0841284886 2694560424 1965285022 2106611863
0674427862 2039194945 0471237137 8696095636 4371917287 4677646575 7396241389 0865832645 9958133904 7802759009

9465764078 9512694683 9835259570 9825822620 5224894077 2671947826 8482601476 9909026401 3639443745 5305068203
4962524517 4939965143 1429809190 6592509372 2169646151 5709858387 4105978859 5977297549 8930161753 9284681382
6868386894 2774155991 8559252459 5395943104 9972524680 8459872736 4469584865 3836736222 6260991246 0805124388
4390451244 1365497627 8079771569 1435997700 1296160894 4169486855 5848406353 4220722258 2848864815 8456028506
0168427394 5226746767 8895252138 5225499546 6672782398 6456596116 3548862305 7745649803 5593634568 1743241125
1507606947 9451096596 0940252288 7971089314 5669136867 2287489405 6010150330 8617928680 9208747609 1782493858
9009714909 6759852613 6554978189 3129784821 6829989487 2265880485 7564014270 4775551323 7964145152 3746234364
5428584447 9526586782 1051141354 7357395231 1342716610 2135969536 2314429524 8493718711 0145765403 5902799344
0374200731 0578539062 1983874478 0847848968 3321445713 8687519435 0643021845 3191048481 0053706146 8067491927
8191197939 9520614196 6342875444 0643745123 7181921799 9839101591 9561814675 1426912397 4894090718 6494231961

5679452080 9514655022 5231603881 9301420937 6213785595 6638937787 0830390697 9207734672 2182562599 6615014215
0306803844 7734549202 6054146659 2520149744 2850732518 6660021324 3408819071 0486331734 6496514539 0579626856
1005508106 6587969981 6357473638 4052571459 1028970641 4011097120 6280439039 7595156771 5770042033 7869936007
2305587631 7635942187 3125147120 5329281918 2618612586 7321579198 4148488291 6447060957 5270695722 0917567116
7229109816 9091528017 3506712748 5832228718 3520935396 5725121083 5791513698 8209144421 0067510334 6711031412
6711136990 8658516398 3150197016 5151168517 1437657618 3515565088 4909989859 9823873455 2833163550 7647918535
8932261854 8963213293 3089857064 2046752590 7091548141 6549859461 6371802709 8199430992 4488957571 2828905923
2332609729 9712084433 5732654893 8239119325 9746366730 5836041428 1388303203 8249037589 8524374417 0291327656
1809377344 4030707469 2112019130 2033038019 7621101100 4492932151 6084244485 9637669838 9522868478 3123552658
2131449576 8572624334 4189303968 6426243410 7732269780 2807318915 4411010446 8232527162 0105265227 2111660396

6655730925 4711055785 3763466820 6531098965 2691862056 4769312570 5863566201 8558100729 3606598764 8611791045
3348850346 1136576867 5324944166 8039626579 7877185560 8455296541 2665408530 6143444318 5867697514 5661406800
7002378776 5913440171 2749470420 5622305389 9456131407 1127000407 8547332699 3908145466 4645880797 2708266830
6343285878 5698305235 8089330657 5740679545 7163775254 2021149557 6158140025 0126228594 1302164715 5097925923
0990796547 3761255176 5675135751 7829666454 7791745011 2996148903 0463994713 2962107340 4375189573 5961458901
9389713111 7904297828 5647503203 1986915140 2870808599 0480109412 1472213179 4764777262 2414254854 5403321571
8530614228 8137585043 0633217518 2979866223 7172159160 7716692547 4873898665 4949450114 6540628433 6639379003
9769265672 1463853067 3609657120 9180763832 7166416274 8888007869 2560290228 4721040317 2118608204 1900042296
6171196377 9213375751 1495950156 6049631862 9472654736 4252308177 0367515906 7350235072 8354056704 0386743513
6222247715 8915049530 9844489333 0963408780 7693259939 7805419341 4473774418 4263129860 8099888687 4132604721

THE FIRST 10,000 DECIMAL PLACES OF π

FROM 5001D TO 10000D

5695162396 5864573021 6315981931 9516735381 2974167729 4786724229 2465436680 0980676928 2382806899 6400482435
4037014163 1496589794 0924323789 6907069779 4223625082 2168895738 3798623001 5937764716 5122893578 6015881617
5578297352 3344604281 5126272037 3431465319 7777416031 9906655418 7639792933 4419521541 3418994854 4473456738
3162499341 9131814809 2777710386 3877343177 2075456545 3220777092 1201905166 0962804909 2636019759 8828161332
3166636528 6193266863 3606273567 6303544776 2803504507 7723554710 5859548702 7908143562 4014517180 6246436267
9456127531 8134078330 3362542327 8394497538 2437205835 3114771199 2606381334 6776879695 9703098339 1307710987
0408591337 4641442822 7726346594 7047458784 7787201927 7152807317 6790770715 7213444730 6057007334 9243693113
8350493163 1284042512 1925651798 0694113528 0131470130 4781643788 5185290928 5452011658 3934196562 1349143415
9562586586 5570552690 4965209858 0338507224 2648293972 8584783163 0577775606 8887644624 8246857926 0395352773
4803048029 0058760758 2510474709 1643961362 6760449256 2742042083 2085661190 6254543372 1315359584 5068772460

2901618766 7952406163 4252257719 5429162991 9306455377 9914037340 4328752628 8896399587 9475729174 6426357455
2540790914 5135711136 9410911939 3251910760 2082520261 8798531887 7058429725 9167781314 9699009019 2116971737
2784768472 6860849003 3770242429 1651300500 5168323364 3503895170 2989392233 4517220138 1280696501 1784408745
1960121228 5993716231 3017114448 4640903890 6449544400 6198690754 8516026327 5052983491 8740786680 8818338510
2283345085 0486082503 9302133219 7155184306 3545500766 8282949304 1377655279 3975175461 3953984683 3936383047
4611996653 8581538420 5685338621 8672523340 2830871123 2827892125 0771262946 3229563989 8989358211 6745627010
2183564622 0134967151 8819097303 8119800497 3407239610 3685406643 1939509790 1906996395 5245300545 0580685501
9567302292 1913933918 5680344903 9820595510 0226353536 1920419947 4553859381 0234395544 9597783779 0237421617
2711172364 3435439478 2218185286 2408514006 6604433258 8856986705 4315470696 5747458550 3323233421 0730154594
0516553790 6866273337 9958511562 5784322988 2737231989 8757141595 7811196358 3300594087 3068121602 8764962867

4460477464 9159950549 7374256269 0104903778 1986835938 1465741268 0492564879 8556145372 3478673303 9046883834
3634655379 4986419270 5638729317 4872332083 7601123029 9113679386 2708943879 9362016295 1541337142 4892830722
0126901475 4668476535 7616477379 4675200490 7571555278 1965362132 3926406160 1363581559 0742202020 3187277605
2772190055 6148425551 8792530343 5139844253 2234157623 3610642506 3904975008 6562710953 5919465897 5141310348
2276930624 7435363256 9160781547 8181152843 6679570611 0861533150 4452127473 9245449454 2368288606 1340841486
3776700961 2071512491 4043027253 8607648236 3414334623 5189757664 5216413767 9690314950 1910857598 4423919862
9164219399 4907236234 6468441173 9403265918 4044378051 3338945257 4239950829 6591228508 5558215725 0310712570
1266830240 2929525220 1187267675 6220415420 5161841634 8475651699 9811614101 0029960783 8690929160 3028840026
9104140792 8862150784 2451670908 7000699282 1206604183 7180653556 7252532567 5328612910 4248776182 5829765157
9598470356 2226293486 0034158722 9805349896 5022629174 8788202734 2092222453 3985626476 6914905562 8425039127

5771028402 7998066365 8254889264 8802545661 0172967026 6407655904 2909945681 5065265305 3718294127 0336931378
5178609040 7086671149 6558343434 7693385781 7113864558 7367812301 4587687126 6034891390 9562009939 3610310291
6161528813 8437909904 2317473363 9480457593 1493140529 7634757481 1935670911 0137751721 0080315590 2485309066
9203767192 2033229094 3346768514 2214477379 3937517034 4366199104 0337511173 5471918550 4644902636 5512816228
8244625759 1633303910 7225383742 1821408835 0865739177 1509682887 4782656995 9957449066 1758344137 5223970968
3408005355 9849175417 3818839994 4697486762 6551658276 5848358845 3142775687 9002909517 0283529716 3445621296
4043523117 6006651012 4120065975 5851276178 5838292041 9748442360 8007193045 7618932349 2292796501 9875187212
7267507981 2554709589 0455635792 1221033346 6974992356 3025494780 2490114195 2123828153 0911407907 3860251522
7429958180 7247162591 6685451333 1239480494 7079119153 2673430282 4418604142 6363954800 0448002670 4962482017
9289647669 7583183271 3142517029 6923488962 7668440323 2609275249 6035799646 9256504936 8183609003 2380929345

9588970695 3653494060 3402166544 3755890045 6328822505 4525564056 4482465151 8754711962 1844396582 5337543885
6909411303 1509526179 3780029741 2076651479 3942590298 9695946995 5657612186 5619673378 6236256125 2163208628
6922210327 4889218654 3648022967 8070576561 5144632046 9279068212 0738837781 4233562823 6089632080 6822246801
2248261177 1858963814 0918390367 3672220888 3215137556 0037279839 4004152970 0287830766 7094447456 0134556417
2543709069 7939612257 1429894671 5435784687 8861444581 2314593571 9849225284 7160504922 1242470141 2147805734
5510500801 9086996033 0276347870 8108175450 1193071412 2339086639 3833952942 5786905076 4310063835 1983438934
1596131854 3475464955 6978103829 3097164651 4384070070 7360411237 3599843452 2516105070 2705623526 6012764848
3084076118 3013052793 2054274628 6540360367 4532865105 7065874882 2569815793 6789766974 2205750596 8344086973
5020141020 6723585020 0724522563 2651341055 9240190274 2162484391 4035998953 5394590944 0704691209 1409387001
2645600162 3742880210 9276457931 0657922955 2498872758 4610126483 6999892256 9596881592 0560010165 5256375678

Facsimile of the first two pages of the computer print-out obtained by Shanks and Wrench, who programmed an IBM 704 to compute π to 100,265 decimal places in July 1961 (see pp. 184-185).